Removing the Barriers to Efficient Manufacturing

Real-World Applications of Lean Productivity

Removing the Barriers to Efficient Manufacturing

Real-World Applications of Lean Productivity

Daniel L. Ferguson

CRC Press is an imprint of the
Taylor & Francis Group, an **informa** business

A PRODUCTIVITY PRESS BOOK

CRC Press
Taylor & Francis Group
6000 Broken Sound Parkway NW, Suite 300
Boca Raton, FL 33487-2742

Printed in the United States of America on acid-free paper
Version Date: 20121120

International Standard Book Number: 978-1-4665-5551-8 (Paperback)

Library of Congress Cataloging-in-Publication Data

Ferguson, Daniel L.
Removing the barriers to efficient manufacturing : real-world applications of lean productivity / Daniel L. Ferguson.
p. cm.
Includes bibliographical references and index.
ISBN 978-1-4665-5551-8
1. Lean manufacturing. 2. Process control. 3. Manufacturing processes. I. Title.

TS155.F4987 2013
658.5--dc23 2012037359

Visit the Taylor & Francis Web site at
http://www.taylorandfrancis.com

and the CRC Press Web site at
http://www.crcpress.com

Dedication

This book is dedicated to all of my grandchildren,
who have brought so much joy into our lives.

Contents

Preface

Where We Are and How We Got Here

When I think of all the evils that have befallen the American economy, I can think of none worse than the concept espoused some years ago that America had "advanced" from a manufacturing-based economy to a service-based economy. At the time, I thought, "This is nuts!" How stupid. How can a nation create wealth by servicing other people's manufactured goods unless we are the only people on the planet who have this magical ability (like the navigators in *Dune* who were the only entities who could "fold space")? Since this is clearly not the case, the entire premise seems ridiculous. Yes, there are some very sophisticated services available for which people are willing to pay handsomely. But, and it is a big fat but, the people who buy these services pay for them with *wealth they already created for themselves*. So, who is the more advanced: the people who can pay to have their yards attended to by a landscaping company or the landscapers? If you said the landscapers, go to the back of the class and stand in the corner.

Those who promoted America's advancement to a service-based economy are at the root of our current economic problems. It is an erroneous mindset that has been infused into our national fabric and has even corrupted the starting point of our economic food chain: our education system. The emphasis for many years has been on building self-esteem, not creating determined, gritty scientists and engineers. If you doubt this, look at the international test results that compare math and science scores for the various age groups. Yes, our 4th graders do OK, but then our kids begin to fall significantly behind by 8th grade, and score near the bottom by age 15 and later.

Some observers say our education system is improving, while others say we have made no appreciable progress for years. In either case, what is the difference? The most recent test results speak for themselves. We are not cutting it, and our children are paying the price. I know many parents who have paid big bucks to send their kids to college for degrees that will not get them jobs that will afford them a standard of living anywhere near that of their parents. Due to the loss of our manufacturing base, many of those really good-paying jobs do not exist anymore. And even if they did, without the proper education, our children would not be filling them.

Regardless of your perspective on our educational progress, it is a fact that American manufacturing has been on the decline for the past two to three decades. We lost our edge years ago. When you couple insufficient education with a national mindset that we no longer need manufacturing, what do you suppose the outcome will be? I will tell you what—it has culminated in the bankruptcy of General Motors and Chrysler, and a multitude of companies before them, leading to an area of the country that we now refer to as the "Rust Belt." Think about it. We invented television but do not manufacture them here any more. We invented large-scale steel manufacturing but now produce less than Japan, less than half of the European Union, and one-fifth that of China. Almost *all* consumer electronics, including personal computers, are manufactured offshore and many of them under original U.S. brand names that were renowned in past decades for quality. The top-rated automobiles are Japanese. Top-rated electronics come from Japan, Korea, and Taiwan. If it were not for Silicon Valley, we could officially declare ourselves a banana republic and plan our future around that instead of continuing to delude ourselves that we are on our way back.

The Change We Need Now

Many people think the key to our future prosperity involves "change." In fact, President Barack Obama was elected on that basis. But, the question is: What kind of change? Without a wealth creation engine of some kind, it would appear that the change we can look forward to is the invention of new and improved methods for divvying up our shrinking economic pie so that it is divided more "fairly." Unfortunately, as we do that, the people who actually own the pie may decide that there is no longer sufficient incentives to justify risking their fortunes in the United States and move what is left of their manufacturing wealth creation enterprise to a country that wants it like, say, India.

The only way you can have a bigger pie, and hence bigger slices for everybody, is if you do something actively to *make* it bigger. It follows then that we simply *must* reinvigorate our manufacturing base. And, since we have not been able to do that in the past couple of decades, there must be some massive barriers out there preventing it. Our task then is to determine what those barriers are and knock them down. This will be a recurring theme throughout this book.

The bottom line is that if you agree that there is a need for real change in how we execute (and value) manufacturing, then you should find what follows useful. On the other hand, if you are satisfied with the way things are going in your manufacturing operation today, then you may already be enlightened and just may be curious to see whether there is anything you have overlooked.

Acknowledgments

I would like to sincerely thank the Statit® Software Company for kindly allowing me to include their excellent summary of statistical methods as Appendix 2 to this book.

Thanks also go to the Lightning Calculator company for allowing me to include their write-up of the red bead experiment as Appendix 1.

I am grateful to my friend and colleague Joseph A. Count for reviewing Chapters 5 and 8. Thank you.

Finally, I would like to thank my wife, Patricia, for proofreading the manuscript and for her enthusiastic encouragement.

Introduction

There was a joke in the 1980s that went something like this:

Three employees of companies and of different nationalities were caught up in a coup in a small, foreign country and scheduled to be executed as spies by the new dictator. Before execution, they were asked if they had any last requests. The first—a Frenchman—said he would like to hear "La Marseillaise" played once more. The second—from Japan—said he would like to give one more lecture on Japanese-style management. The third—an American—said he would like to be shot first so he would not have to listen to another lecture on Japanese-style management.

For decades, the United States has been enamored with "Japanese-style management" and tried somewhat to replicate it in the United States. And with good reason—we all know that the Japanese have been very successful at producing quality, high-value products since the 1960s. What many Americans do not know is that an American scientist and statistician, W. Edwards Deming, is largely responsible for Japan's success. Yet, if you ask someone on the street who he is, they have probably never heard of him. That is because Dr. Deming's methodology never completely caught on in the United States, which is unfortunate because the United States has never completely caught up with the Japanese in manufacturing.

Deming's central premise is that improving product quality will increase productivity and competitive position and ensure long-term survival of the company. His methodology for achieving higher quality levels is based on his landmark 14 Points, and Point 12 is the inspiration for this book. It says that management's job is to remove the barriers that keep people from taking pride in their work, which is a profoundly different perspective than others, which say that it is management's job to motivate their employees.

Deming directed his 14 Points specifically to the highest levels of management, who are expected to handle the parts they are responsible for, like product design and marketing, while pushing the manufacturing parts down to their plant managers. For the most part, corporate America has not embraced Deming's principles, so the question is can they be applied at the plant level without direction from senior management? My answer is a definite Yes! and this book is an attempt to show the reader how. The one qualifier is that at the plant level there is usually little or no input to the design of the finished product, so the work is focused on reduction of variability and elimination of waste and delay in the manufacturing process.

Since I cannot improve on Dr. Deming's principles, and have verified them through personal experience as being "truth," the 14 Points are the foundation for my "Model Vision" of what a well-run manufacturing plant looks like. The book provides all the steps in the order needed to achieve that vision and overcome the barriers encountered in the typical manufacturing plant. Once these barriers are removed, the people who work in the plant are freed to achieve the highest levels of quality and productivity capable in the system.

Removing barriers is just the beginning. It does not stop there. System improvement must continue even after every known barrier is removed to drive variability lower and lower. The organizational structure and methods presented here are designed specifically to do that.

Each chapter contains essential elements of efficient manufacturing in sufficient detail for the reader to gain understanding of the concepts. However, volumes have been written on many of them, and this book is a road map, not a detailed treatise, so you may wish to dig deeper into some of the topics. Concepts aside, nothing can be accomplished without specific actions of the people in the various manufacturing functions. Where this book differs from other works is that it outlines specific *activities* management must be engaged in to eliminate the barriers to producing products that are consistently defect free. While much of this is dependent on processes and equipment, organizational relationships at all levels also play a major role and are discussed in detail.

The methods presented here are proven to be effective. To provide authenticity, specific real-life examples are sprinkled throughout, and the book closes with a case study to show how the process can work if given a chance.

This book was written specifically for manufacturing plant managers and their staff—it is hoped people who love their products and the people who produce them. Instead of being on a "random walk," it provides each manufacturing functional group with straightforward directions for creating a smooth-running facility where reducing variability is the order of the day, and everyone is focused on improving the system. It is far from theoretical. In fact, it is the opposite—one might even call it a "cookbook." All the reader has to do is complete the "work assignments" given in each chapter, in the order presented, and the result should be achievement of the Model Vision for a manufacturing facility. This effort will require dedication by all involved. This is *not* a "program-of-the-month" approach, and for some, it will be a new way of living. But, the end result will be worth the effort in terms of financial and personal gain and may just keep you in business for the next 100 years.

Good luck!

Dan Ferguson

About the Author

Daniel L. Ferguson received a BS (1967) and MS (1972) in mechanical engineering with a major in automatic controls from Clemson University.

After receiving a direct commission in 1969, he served as a U.S. Army aircraft maintenance officer in Vietnam from 1970 to 1971.

Over his 40-year career in industry, he has held engineering and management positions in operations and maintenance with two Fortune 500 companies, between which, in the early 1980s, he managed a polyester resins plant for one of the major producers.

While serving as facilities and maintenance manager for a large manufacturing facility, his plant was tapped to be the pilot site for implementation of SAP plant maintenance for the corporation. (SAP is a well-known Enterprise Resource Planning [ERP] software system.) As part of the SAP project team, he developed a number of systems and plant structures that are still in use today.

After a successful pilot, he was assigned to the SAP implementation team and was responsible for rollout at similar manufacturing facilities. Later, this responsibility was expanded to implementation at new acquisition companies. Considered an expert in this area, he was asked to resume this work on a consulting basis after retirement in 2008.

As a lifelong audiophile, he has written three books and a number of articles on loudspeakers and electronics. His current project of interest is development of a new orthopedic device to aid patients with Parkinson's disease.

He has three grown children and currently resides with his wife of 46 years in Appleton, Wisconsin.

Chapter 1

Deming Got It Right

Introduction

Back in the late 1970s, when I first started in management, I became increasingly frustrated with the lack of discipline I encountered in the manufacturing workplace. We made many errors that resulted in lost productivity and sometimes marginal product quality. The system relied heavily on luck. This was because the processes were largely out of control and produced considerable variability in the finished products. Added to that was frequent equipment breakdowns, some of which were near catastrophic. Workers were inadequately trained, and some of them were in positions for which they were unqualified. I even encountered some finished product designs that appeared to be noncompetitive because of the inferior raw materials used in order to reduce costs.

On my own, I set out to improve the processes I was responsible for and even met with some successes, but I continued to have negative feelings about the organizations I worked for and their overall philosophies. Around 1984, I had the opportunity to see Dr. W. Edwards Deming's videotape series. All I can say is that it was *life changing*. In the first couple of sessions, I felt completely validated as a person. He presented his methodology for improved quality and productivity in practical, real-life settings that reinforced my own experiences and made my personal dissatisfaction seem logical and rational. But most of all, Dr. Deming gave me a confident, clear path forward.

At this point, it might be helpful to provide a little background on who Dr. Deming is and why his work is so significant. Born in 1900, he lived until 1993. He earned a BS in electrical engineering from the University of

Wyoming, an MS from the University of Colorado, and a PhD from Yale. Both graduate degrees were in math and physics. He was a statistician, professor, lecturer, and international consultant. But first and foremost, he was a scientist. In his video series, he likes to refer to himself as "an apprentice statistician" because his quality improvement process relies heavily on statistical methods. Without them, he emphasizes that it would be nearly impossible to arrive at the right corrective action. Having said that, the Deming process consists of 14 points, only two of which specifically refer to the use of statistical methods. The rest are practical, commonsense essentials that once you have seen, you have a tendency to smack yourself in the forehead and let out a big Homer Simpson "Doh!" If you are not working every day on the 14 Points, he says you are on a "random walk." And to that, I wholeheartedly agree.

Dr. Deming has been credited with improving production efficiency in the United States during World War II, but he is undoubtedly best known for his work in Japan. Whereas for many years he received only limited recognition in the United States, the Japanese in today's vernacular "got him" (as in they understood, appreciated, and even venerated him). When I was a kid in the 1950s, "made in Japan" was synonymous with "junk." Any Japanese toy I had was a cheap facsimile of the real thing and tended to fall apart. Who in their right mind would say that about Japanese products today? The change from cheap junk to quality and value is the direct result of implementing Deming's methods. Instituted in 1950, the top companies in Japan are, to this day, awarded the Deming Prize.

Deming's main thesis is that if you improve the quality of your product, you will, as an inherent consequence, improve both productivity *and* competitive position, hence the title of his landmark book, *Quality, Productivity, and Competitive Position* published by MIT Press in 1982. The book was later retitled *Out of the Crisis.* I had the opportunity to put it to work for some years and can attest to the validity of the premise. For some unexplained reason, it has still not "caught on" in the United States even though there are many Deming disciples out there. Unfortunately, those disciples do not appear to be in top management. If they were, we would not have lost all those millions of manufacturing jobs.

So, what then is the purpose of this book? In short, it is to provide readers with a clear path for improving their manufacturing processes to enable them to stay in business for as long as they wish. Along the way, personal experiences and observations will be intertwined with Dr. Deming's road map to bring it to life. Since the basis for everything that follows is the 14 Points, here they are with permission from MIT Press:

Deming's 14 Points for Top Management (Copyright 1982)

1. Create constancy of purpose toward improvement of product and service, with a plan to become competitive and stay in business. Decide whom top management is responsible to.
2. Adopt the new philosophy. We are in a new economic age. We can no longer live with commonly accepted levels of delays, mistakes, defective materials, and defective workmanship.
3. Cease dependence on mass inspection. Require, instead, statistical evidence that quality is built in, to eliminate the need for inspection on a mass basis. Purchasing managers have a new job, and must learn it.
4. End the practice of awarding business on the basis of price tag. Instead, depend on meaningful measures of quality, along with price. Eliminate suppliers that cannot qualify with statistical evidence of quality.
5. Find problems. It is management's job to work continually on the system (design, incoming materials, composition of material, maintenance, improvement of machine, training, supervision, retraining).
6. Institute modern methods of training on the job.
7. Institute modern methods of supervision of production workers. The responsibility of foremen must be changed from sheer numbers to quality. Improvement of quality will automatically improve productivity. Management must prepare to take immediate action on reports from foremen concerning barriers such as inherited defects, machines not maintained, poor tools, fuzzy operational definitions.
8. Drive out fear, so that everyone may work effectively for the company.
9. Break down barriers between departments. People in research, design, sales, and production must work as a team to foresee problems that may be encountered with various materials and specifications.
10. Eliminate numerical goals, posters, and slogans for the work force asking for new levels of productivity without providing methods.
11. Eliminate work standards that prescribe numerical quotas.
12. Remove barriers that stand between the hourly worker and his right to pride of workmanship.

13. Institute a vigorous program of education and retraining.
14. Create a structure in top management that will push every day on the above 13 points.

Please take a few minutes and study these carefully as each one is significant. After that, I provide a few comments on each one before continuing.

Point 1: Management Must Have "Constancy of Purpose" to Stay in Business

This seems obvious doesn't it? And yet, how many products do we bring home that we are truly satisfied with? How many do we regard as just plain junk? How many were defective and had to be returned for replacement? How many cars have you owned in your lifetime that you were truly happy with? In the extreme, you may even ask yourself, "What could they have possibly been thinking?" So, does the management of the companies that make inferior products have constancy of purpose? Is management really planning to stay in business for the foreseeable future with these junky products?

For a company to ensure that it stays in business, its products must "delight the customer" by giving them *more* than they expected. If the design and function of the product are exceptional, it will not matter unless the product is manufactured without defects. Only top management can direct this as, sometimes, it will result in reduced short-term profits in order to be able to invest in the company. What chief executive officer (CEO) is willing to do that when it seems that all many of them can do to cut costs is by eliminating jobs to make the quarterly earnings report "meet expectations" on Wall Street?

Point 2: Adopt the New Philosophy

To survive in today's competitive world, we have to be extremely good at what we do. We are pitting our skills and technology against an emerging global labor force that has a wage rate that is a tenth of the United States and is able to do a lot with very little in the way of modern machinery. Therefore, we have to be 10 times more productive to compete on an even basis. You cannot do that if you are making mistakes that generate scrap and rework. The new philosophy says that we will do whatever it takes to get it right the first time.

Point 3: Cease Dependence on Mass Inspection

Deming says you cannot inspect quality into a product. It is built in and merely mirrors the quality of the process that produced it. So, even if you perform a so-called 100% inspection, defects will still get through. The alternative is rigorous, statistical sampling that measures the critical parameters that *characterize* the quality of the product. Your task is to drive those measurements to more and more favorable levels.

Point 4: Stop Doing Business with the Low Bidder

It goes without saying that consistent, high-quality raw materials are the starting point or they will become a source of defects. There are no "bargains" out there when it comes to raw materials. You cannot afford bargains. You can only *afford* quality goods (and services) from suppliers who are business partners. Several real-life examples are given in further discussion to show how dramatically this works.

Point 5: Find Problems

There is some goofy notion that workers are not only responsible for reporting problems but also responsible for trying to fix them. (Isn't that why quality circles were invented?) The truth is that it is *management's* job to find *and fix* problems. Yes, the workers are participants and often are the best source of information. However, what they observe on the job may be far removed from the root cause of the problems. While workers must be free to contribute, it is engineers and technicians who have the responsibility to *create* lasting solutions. Their creativity can only be unleashed by a supportive and involved management team.

Point 6: Institute Modern Training Methods

The best investment you can make is in your workforce. Machines are stupid and only do exactly what they are told. Imagine what your business would be like if your employees operated in that mode because they did not have enough training to be able to look ahead and advise management of potential train wrecks. Modern training methods require certified instruction and training materials along with testing to *measure* how much knowledge has been transferred. If your training program consists entirely of having a new employee "bird dog" a more senior employee on the job, then that would be the metaphysical opposite of modern.

Point 7: Institute Modern Methods of Supervision

Each level of supervision must be marked by technical competence so that each level is a resource for the ones below it. If your organization has operational layers that are management training positions, you will have gaps and inconsistencies. What do these trainees know about the jobs of those reporting to them? How will they be able to help anyone? Who will the workforce go to for assistance?

Point 8: Drive Out Fear

An organization that is free of fear is liberated. All employees must be free to tell the truth without fear of retribution. To do that they must be made to feel that management is their *advocate*—not enforcer. Management is who you go to in order to get your problem solved.

Point 9: Break Down Barriers between Departments

Let us face it. Everywhere you go, you can find territorial people who think their function is the most important. There are three basic and equal manufacturing functions: the doers (operations), those who produce the product; the fixers (maintenance, production engineering), those who keep the process running; and the thinkers (technical, process engineering, research and development), those who reengineer everything to make it better. It takes leadership by top management to mold these disparate types of people into a team. (The overused, clichéd word *team* is expanded in further discussion.)

Point 10: Eliminate Goals, Posters, and Slogans Directed at Workers to Do More

Have you ever worked in a place that displayed big banners like, "Goal: Increase Productivity by 10%" or "Be Safe," or the like? Why not just put up a banner that says, "Pay Attention and Do Your Job Right Today for a Change"? Deming has a lot to say about this subject, so we spend some time on it in further discussion.

Point 11: Eliminate Numerical Quotas

Work standards are either too high or too low. When I was in college, I worked at a place in the summer that had them, and I can tell you that they are stupid. People should be free to make as much production as the system is capable of producing. The system will determine that amount from day to day depending on the health of the process. Trying to coerce workers into pushing beyond that is a recipe for quality and safety problems. What this boils down to is a matter of trust. Setting numerical quotas tells the employees that you do not trust them to make all the quality product they can. You are assuming they are lazy and will only do enough to keep from being fired.

Point 12: Remove the Barriers That Keep People from Taking Pride in Their Work

In my opinion, this one point is the most profound of all and encompasses almost *all* of the other points. The wording is nothing short of inspired. What it means is the vast majority of people *want* to do a good job because taking pride in one's work is *natural*. People have this desire built in. Mismanagement of people, equipment, raw materials, and processes causes defects that are barriers to that natural human response. Removing those barriers changes *everything* for the people doing the work and unleashes the power of the human spirit.

Point 13: Institute Vigorous Education and Retraining

Do not assume that everyone in your organization has sufficient process knowledge or even enough verbal and math skills to do a top-notch job. Also, people learn things at different rates. It is management's job to *measure* what people know and train and retrain them as necessary for them to be successful in their jobs. Insufficient training is just another barrier that can keep people from doing their jobs well enough to take pride in them.

Point 14: Top Management Must "Push" (Measure and Report) Every Day on the Above 13 Points

Quality improvement, according to Deming, is a never-ending, iterative process. All of the preceding 13 points have to be worked on every day by management, and progress *must be measured and reported.* These points are not the program of the month. In the context of manufacturing, they are a new way of living.

Before going any further, it must emphasized that this feeble attempt to summarize W. Edwards Deming's 14 Points is just that—a feeble attempt. To fully understand their true depth and significance, you must go to the source, his book, *Quality, Productivity, and Competitive Position,* or the later version, *Out of the Crisis,* and spend a month or so studying these points in detail along with the other invaluable information you will find there. It is also recommended that you get a copy of his video lecture series and watch it with your team. Then, you might be ready to jump in.

Again, the 14 Points are the foundation we will be building on in our quest for better manufacturing. If you have doubts about their validity, then read no further. If, on the other hand, you see their wisdom, let us move on.

The Vision

To create a manufacturing process that is better than the one you have today, you will need "a vision." Yes, this term evokes all kinds of Dilbert-type responses, like "Serving humankind through innovation" and the like. While not a vision statement, a well-known chemical company had this slogan in the 1960s: "Without chemicals, life itself would be impossible." Dilbert would be proud of that one.

The point is that a vision for your plant, operation, or process should be something you can share with your employees and not look stupid. So sit back, prop your feet up on your desk, and think what your life (and the lives of your employees) would be like if

1. Your workplace was clean, well lit, properly ventilated, and free of safety hazards.
2. Your machines were in tip-top mechanical condition and capable of running without breakdown from one maintenance session to the next.
3. Your process was monitored and controlled with modern instrumentation.
4. You used only high-quality raw materials.
5. Your workforce was trained to be "highly skilled professionals."
6. Your operators monitored product characteristics using standardized methods and adjusted settings using statistical process control techniques.
7. Product quality was measured and reported using appropriate statistical methods.
8. Each day, cross-functional teams analyzed the process and product data and looked for improvement opportunities.
9. Production schedules were easily maintained because surprise quality defects were nonexistent, and your operation is equipped to make changeovers rapidly.
10. Your employees had the freedom from fear to try to make things better.

You probably have other characteristics you want to add to this list, but for the sake of discussion, for now we call this our "Model Vision." To be considered credible in today's business environment, we also need a "vision statement" (I say, tongue in cheek). So, what would a corresponding vision statement look like? How about, "Removing the barriers that keep people from taking pride in their work"? OK, obviously I stole that from Dr. Deming. However, if you think it is trite, try mentally replacing each positive characteristic with its opposite. For example, your workplace is dirty, your machines are falling apart, your process is out of control, and so on—*bizarro world*. Then, put yourself in the place of the employees who have to work in this negative environment. Just how much pride will they take in their daily grind to try to keep things patched up, knowing that they are making defects that will eventually cost them their jobs?

So, are these the right potential barriers or not? If they are, removing them is management's primary job. In other words, it is *your* job.

The next question is: How does your facility measure up at this point in time? Does the model vision represent one you would aspire to? If it is, and you know you have opportunity areas, are you actually willing to try to do something about it?

Reducing Variability Is the Key

In the movie *City Slickers*, the old cowboy, Curly (played by Jack Palance), told city slicker Billy Crystal that there was only one thing important in life—"just one thing." When Billy Crystal asked what that was, Curly said that he would have to figure that out for himself. As it turns out, Curly was right. When it is all said and done, the single thing you have to work on to make your entire manufacturing operation roar to life is *reducing variability in your finished product.* It is that simple (and that complex).

Dr. Deming illustrated this concept elegantly with his famous Red Bead Experiment. He devised a paddle with 50 drilled holes of a certain size that would retain plastic beads when dipped into a bowl full of them. The bowl contained an 80/20 mix of white and red beads—white symbolizing good product and red representing defects. As one would expect, when a person dipped the paddle into the container, stirred the contents, and withdrew the paddle, it retained a representative mix of colors. After repeating the process a number of times and plotting the red bead count on a chart, the process generated a classic normal distribution curve with a mean near 10, the median of the known concentration of red beads.

The experiment clearly illustrates that systems that are "in statistical control" obey the laws of randomness and can be accurately characterized using statistical methods. The quantity of red beads you get out is simply representative of the system—no more, no less. Thus, if you want a process that produces fewer defective products, you have to "get the red beads out of the system." You do this by reducing variability at every step in the process.

The Red Bead Experiment illustrates a number of real-world lessons that are integral to Dr. Deming's 14 Points, and he used it in his video and lectures. Lightning Calculator has developed an excellent write-up and made it available on the Internet. For convenience, it is included as Appendix 1 with the kind permission of Lightning Calculator. It is recommended that you study it thoroughly and try it out with your team. It is both instructional and fun, and kits are available on their Web site (http://www.qualitytng.com/).

Now look back at our 10-point Model Vision. I hope the points are self-evident by now, and you agree that each one is essential and in the correct order. This is crucial because every one of them will have a major impact on reducing variability.

Chapter 2

Removing Barriers in the Workplace

Introduction

One of the most rewarding things we can do in life is a "good day's work." While this may sound corny, some of us have had the opportunity to see firsthand how happy people look when they leave the job after a productive day, when everything has gone right. It gets right back to Deming's Point 12: People want to take pride in their work. The problem is that the barriers they have to deal with each day can make that impossible.

So, let us start with the easiest potential barrier first: the workplace. We said in Point 1 of our Model Vision that we wanted ours to be clean, well lit, properly ventilated, and free of safety hazards. Let us talk about each of them.

A clean workplace is essential for a number of reasons.

- It reduces the chances of product contamination.
- There will be fewer mistakes from using the wrong tools and materials.
- The equipment will run more reliably.
- It conveys the message that the company has high standards for itself and its employees.
- It is safer to work in a clean, uncluttered environment.
- It instills pride in the employees.
- It is healthier.
- If you allow customers to visit your plant, they will like what they see.
- You may be able to negotiate lower insurance rates.

Feel free to add to the list, but these will do for starters.

Order and Cleanliness

Order and cleanliness are the necessary starting point for everything that follows, so it is imperative to set the tone for how you plan to conduct business. Understand that this new way of operating will require dedication and commitment by you. Order and cleanliness require daily attention if you are to maintain your standards, which must be done by patient coaching—not complaining.

If your organization did not place much emphasis on housekeeping in the past, then the new standards must be presented at crew meetings so that everyone clearly understands the new expectations. Each person should be given a printed handout for reference. Because many housekeeping problems are beyond the capability of the workers to correct, it is important to encourage team members to point out housekeeping items that need management's involvement. Of course, once they do that, management must deliver and not just pay lip service or this becomes another meaningless exercise.

After the new housekeeping standards have been communicated, managers must walk their departments with area supervisors. Action lists must be generated and funds and resources assigned to upgrade the facilities and equipment. This is not a one-shot activity. The list must be managed on an ongoing basis to ensure improvement. Anything less and this becomes another program of the month.

Progress must be communicated in monthly reports to the organization and top management. You will also need to post progress reports on employee bulletin boards; it forces you either to make progress or to look stupid. Also, once a month, top management must make announced, walking inspections of selected areas to show employees that they are involved and supportive. Praise for progress is essential.

Getting Started

In general, anything ugly should be addressed. Here is a list of obvious things to get you started.

- Floors: These should be clean and free of dust and debris.
- Walls: Walls should be clean and painted a light color, preferably white.
- Lighting: Increased lighting has been shown in past studies to improve morale and productivity, and it allows people to see defects more easily. Also, keeping your building relamped sends a clear message that you

are maintaining your own standards and not just imposing them on the workforce. Feel free to use the most energy-efficient products but keep your lighting levels at known industry standards.

- No leaks: Leaks of any kind are not permitted on an ongoing basis. Known leaks must be barricaded until repairs can be made to prevent slips and falls.
- Storage: This should be designed to house only authorized tools and supplies. Cabinets must be *routinely inspected* to ensure they remain clean and contain no extraneous items. Most of the cabinets I have seen in the past had a tendency to become filthy and full of junk.
- Absence of clutter: There should be nothing found in the workplace that is not authorized on a list that is clearly posted. This is not an easy thing to do since clutter seems to accumulate exponentially and spontaneously.
- Signage: Only professionally done signs and labels are permitted, such as engraved plastic or those that are commercially manufactured. Handmade stickers, tape, and the like are not allowed. For emergency information, use commercially available caution, danger, or lockout tags on a *temporary* basis.
- Bulletin boards: These boards must be under glass and locked and must be reviewed at least weekly to make sure that everything in there is current.
- Process equipment: Sufficient production downtime must be allocated to clean process equipment as necessary. The nature of the process will dictate how much and how often, but it is a cost of doing business. A dirty process produces defects. On the other hand, if your people are spending too much time cleaning up, you probably need to reengineer something.

Whenever possible, the employees who work in an area should do the cleanup because it instills a sense of ownership. When this is not practical, call in the contractors. But in general, teach people that (just like at home) we are each responsible for cleaning up our own messes. From time to time, you may even want to schedule a cleanup day on a weekend or off day. Make it an event and provide food and overtime pay. I did this once at a large plant, and we hauled off seven tractor trailer loads of junk that had accumulated *inside the buildings*.

Ergonomics Are Economical

It goes without saying that each workstation in your plant should be as ergonomic and comfortable as possible. Your employees will do a better job if they are less fatigued and their feet are not killing them from standing *unnecessarily* all day. While many jobs can only be done from a standing position, those that can be done safely from a seated position should be engineered accordingly by providing the correct stools, chairs, and workstands. Whenever possible, let the employees tell you what they need and then provide them *exactly* what they asked for. Be prepared to make adjustments until you get it right.

Employee Facilities Show You Care (or Not)

Clean, modern restrooms and break rooms are mandatory if you expect your employees to take the "new you" seriously. Restrooms, in particular, must be maintained as clean as the ones in your home and provide similar privacy. So, if your restrooms are shoddy, spend some money and renovate them as soon as possible. Nothing fancy is needed. Fresh and clean will do the job nicely. The same is true for break rooms. And, make sure they are both fully stocked at all times with consumable items. Providing quality employee facilities is an opportunity to tell your people you care about them.

Making Safety Equal to Everything Else

Now that the place is cleaned up, the stage is set for fewer accidents. However, unsafe work practices can interfere with that because the vast majority (90% or so) of accidents are caused by unsafe acts, not unsafe conditions. And in fact, the creation of an unsafe condition is usually the result of an unsafe act by someone. That includes the engineers and managers who designed the workplace or neglected to address a hazard that developed over time.

The ability to recognize unsafe acts and conditions does not come naturally. It requires some training, and the easiest and most cost-effective way I have found to become a trained safety observer is the DuPont STOP® program. I learned about it firsthand when I was an employee in the 1970s. There were several bedrock principles that were new to me at the time that just made sense but one stood out above all the others.

Principle number one was that DuPont's goal was to make safety equal to cost, productivity, and quality. This honesty stands in stark contrast to all those phony slogans like "Safety first" or "Safety is our number one priority" because in reality safety is *not* first. Maintaining an enterprise (constancy of purpose) is first. Safety is an *essential component* of that enterprise. You have to do it *all* well if you expect to stay in business.

DuPont determined decades ago that the problem with safety is that, in practice, it actually gets *less* priority than the other objectives. People get in a hurry and do dumb things when they are under pressure to get the work out or deal with problems that cause an interruption in production. Mentally, they know their livelihood depends on making products, and it is hard to turn that off. Being safe in the face of adversity is a learned behavior for many of us.

Another aspect of safety is the design of the process. Cutting corners in equipment design or layout can lead to less-safe operating conditions, which is an unsafe act by the person who made that decision. Merely making safety *equal* to cost will keep that from happening.

If managers spent the same amount of time on safety as they did on operations, paperwork, and meetings, there would be quantum reductions in workplace accidents and injuries. As it is, it is doubtful many of them spend 1% of their time on it, and even then it is usually to participate in an accident investigation. It would be far more effective if managers spent the 1% observing their people working.

DuPont's STOP program is a relatively easy way to train managers and employees to become trained safety observers. It teaches you the common unsafe acts and conditions to look for and how to interact with people in a nonthreatening manner to get them corrected. It also gives you a simple scorecard system you can use to measure your progress. It is highly recommended, but only after you have cleaned up your plant and have it running smoothly. Then, when things have settled down, you can STOP and take a few minutes to focus on the way people work and the potential hazards that have been overlooked in the past.

Establishing Minimum Standards

The STOP program emphasizes a second almost equally important principle. When it comes to safety and housekeeping, *people only meet your minimum standards*. This may sound overly negative or even harsh, but it happens to be true. When it comes to productivity, people have a real-time scorecard to refer to, and they usually try to beat their own expectations. These same people can become annoyed when they are doing an outstanding job in those areas and you come along and point out some trash or clutter in their work area.

Establishing and then maintaining your minimum standards is a process. You establish your standards by walking around with your team leaders and "finger-pointing" the activities of people and housekeeping conditions that do not meet your standards. You repeat this process formally, over time, until your minimum standards are established. Then, each day, as you walk through your area of responsibility, you maintain your standards by picking up incidental trash on the floor you happen to see or pointing out a bigger problem to supervision. The day you walk by a piece of trash on the floor and do nothing is the day you establish a new standard that trash on the floor is OK. That is the principle.

Again, it is extremely important for you to praise and reward sustained safety and housekeeping performance. Show genuine appreciation when the workplace is especially clean and orderly. Housekeeping and safety prizes are an inexpensive way to do that.

Other Safety Requirements

No facility can be considered safe without a formal locking and tagging procedure. Catastrophic injuries and deaths can result from inadvertent energizing of equipment, so this one is not an option. The same is true of confined space entry. Every person at your facility must be trained and retrained on these annually at a minimum. Failure to do this can even put your facility and company in jeopardy and would be the opposite of constancy of purpose.

If you do not have a trained safety professional as a member of your staff, you should consider either hiring one or retaining a consultant to get you started. Safety is an essential vision element. Your operation will never make progress as long as people are being injured. In addition to the direct human losses in pain and suffering, the organization suffers. No matter how much money you invest or how clean and shiny the facility is, confidence in your leadership will be destroyed along with morale.

I have one last thought on safety. Until and unless you get your process running reliably, do not expect a reduction in safety incidents. Breakdowns and process interruptions cause stress in your workforce like nothing else, and people start cutting corners to try to make up lost time. The best thing you can possibly do to improve safety is to find and fix the process problems that are barriers to continuous operation.

Summary

Every facet of your facility should inspire confidence from your workforce. It should display the content of management's character: competent, ethical, and concerned for the people who depend on them. It is a place where individuals become team members.

Chapter 3

Removing the Equipment Reliability Barrier with Effective Maintenance

Introduction

Point 2 in our Model Vision is that our machines must be in tip-top mechanical condition and capable of running without breakdown from one maintenance downtime to the next. There are many detailed treatises available on all the intricacies of plant maintenance because it is a big subject. Volumes have literally been written *and are still being written*. However, this chapter is only intended to give you an overview with enough information for you to select the right maintenance *strategy*. After that, you can dig into the details on your own or even bring in a qualified consultant to help with the "mechanics." Also included are a few lessons learned (some painfully) that should help you avoid some common pitfalls.

To begin, there is nothing that will fill your system with red beads faster than machine breakdowns. In addition to the lost productivity and expense associated with the actual repairs, machines do not typically produce good product one instant and then just stop running. Instead, they limp along on life support while people scramble to figure out whether to shut down or call in assistance in the middle of the night. During those times, quality is suffering along with the machine. Then, after the repairs are made, there is another transition period during startup and final adjustment when quality is again questionable.

The implications of equipment breakdowns are universally understood. For example, here is an *actual* vision statement that was used on a multimillion-dollar maintenance initiative by a large corporation:

> Our Machines Don't Break Down

If only that were true. Instead, many of us struggle daily to keep our machines running from one production batch to the next, hoping to get through with a minimum number of interruptions, all of which introduce variability. If you want to get beyond this, you simply *must* make the process more reliable, and the first step is a comprehensive maintenance plan.

To start with some basics, there are three well-known maintenance strategies:

1. Preventive maintenance (PM): This consists of replacing parts at established intervals that are shorter than expected failure intervals and performing recurring services like lubrication and adjustment. The strategy is to plan the jobs and then perform the work during scheduled downtimes.
2. Predictive maintenance (PdM): PdM consists of monitoring the *condition* of the process *while it is running* and then planning the indicated repairs to be performed during scheduled downtimes (or sooner if the conditions deteriorate).
3. Total productive or operator-assisted maintenance (TPM): Operators are trained to perform limited maintenance tasks to prevent small problems from becoming bigger, like making adjustments and replacing consumable parts. They also assist technicians during maintenance downtimes or whenever needed.

We will need all three.

Preventive Maintenance

Preventive maintenance (PM) is the maintenance most familiar to us. Obviously, there are recurring maintenance tasks that must be done at specific intervals to prevent failures and extend the life of any type of equipment. Filters have finite lifetimes, lubricants break down with heat or become contaminated, and some machine parts wear from the moment they are placed in service and will only last a certain number of hours. Failure to do the *required* PM tasks will have significant consequences.

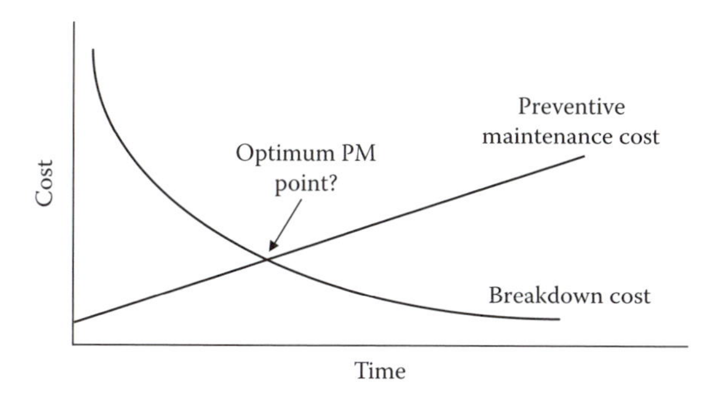

Figure 3.1 Cost of maintenance.

Because PM is expensive, each task should be reviewed periodically to be sure it is essential and that the intervals are realistic. Unfortunately, preventive tasks can become almost a tradition, and their elimination may be resisted if it represents a loss of work to the maintenance workforce. (According to Dr. Harry Levinson, noted psychiatrist, management consultant, and lecturer, "all change involves loss and all loss must be mourned.") To be as objective as possible about determining what is essential and what is overkill, you will need to involve all major functions—maintenance, engineering, and operations—in the process to arrive at a PM system that is cost effective.

Speaking of cost-effectiveness, Figure 3.1 shows the relationship between cost of PM and the cost of breakdowns. If you overmaintain, you have low breakdown costs but high maintenance costs. Conversely, if you undermaintain, you have low maintenance costs but high breakdown costs. There is a sweet spot where the two intersect, and you reach a minimum cost for the two combined. Over time, you will need to keep pestering this trade-off until you are satisfied that you have done all you can to reach the optimum point at which your process is reliable *enough* so that product variability remains in control. In my opinion, that is the only rational optimum.

Predictive Maintenance

Where PM is "old school," predictive maintenance (PdM) is very much high tech. It relies on expensive sensors and analytical software to measure the health of your plant equipment and calculate how much time you have left before you can expect failure.

Here are seven PdM elements, all of which can be very effective:

1. Vibration analysis
2. Dynamic balancing
3. Laser alignment
4. Ultrasonic leak detection
5. Infrared imaging
6. Motor analysis
7. Oil analysis

Vibration Analysis

Vibration analysis involves placing accelerometers or velocity transducers at various points (usually bearings) throughout your process and capturing the vibration spectrum at each of them. The transducers can be permanently mounted at inaccessible locations or portable and handheld and moved from point to point on prescribed routes. The recorded data are downloaded into a personal computer using proprietary (meaning "expensive") software that analyzes each vibration signature and then provides guidance regarding their health. Also, after successive downloads, each new data set is compared to those in history, and the program determines if there are any worsening trends developing. The ultimate application of vibration monitoring is a permanently installed system with continuous, real-time monitoring. This is routinely used on modern high-speed paper machines.

Of all of the PdM tools, vibration analysis probably provides the biggest bang for the buck as it provides the most information. It is capable of showing you problems like imbalance, misalignment, imminent bearing failure, structural resonance points, and even some motor problems. You can either buy the equipment and train a few of your technicians to do the monitoring or you can contract it all out and just get the reports. In either case, vibration monitoring is essential and can drastically reduce or eliminate equipment failures.

There is a certain satisfaction from making rough-running equipment run smoothly. I was on a plant visit one day and one of the lead technicians who knew I had an interest in vibration analysis took me around and had me place my hand on one piece of equipment after another. Each one was so vibration free it was hard to tell it was actually running. The technician had worked for the better part of a year to bring that about, and he was quite proud of what he had accomplished—as well he should have been. Reliability at that facility was greatly improved by his efforts.

Dynamic Balancing

Imbalanced equipment vibrates and eventually self-destructs. In my opinion, it is unnecessary because imbalance is the easiest to detect with vibration analysis and is often easy to correct. The remedy is dynamic balancing using electronic equipment similar to a wheel balancer like the ones at a tire dealer, except more sophisticated and on a larger scale. This type of balancing is best done during equipment rebuilds, when the rotating elements can be removed and placed in the balancing machine. When this is not possible, large components can usually be balanced in place, but it is much more difficult and time consuming, and the results may not be quite as good.

Not all imbalances are weight related. You can get similar vibration readings from asymmetrical fan blades, pump impellers, and electric motor rotors. You just have to keep digging until the source is finally determined.

Laser Alignment

Another common cause of vibration, metal fatigue, and failure is misalignment between mating shafts like motors to pumps or motors to gearboxes. While a skilled millwright can do a pretty fair alignment job with straightedges, squares, and dial indicators, nothing compares to a laser for accuracy. You will want to add this to your PdM arsenal. Laser-aligned equipment should run much smoother and have less bearing and coupling wear than equipment aligned without this benefit.

PdM equipment manufacturers have reasonably priced laser alignment kits available for purchase. However, performing laser alignment requires a fair amount of training, so along with the kit, you will need to invest in training several technicians. Last, it is important to take vibration readings before and after alignment to see if there has been any improvement.

Ultrasonic Leak Detection

High-pressure air and steam leaks produce noise that contains high-frequency components that are beyond the range of human hearing (hence ultrasonic). By using a sensor designed for that range, leaks can be detected even in the noisiest factory environment.

Let us face it, leaks are expensive—especially air leaks. In this day of energy conservation, eliminating leaks is one of the most cost-effective things you can possibly do. As an example, at a large facility, we were able to reduce power consumption by a half a million dollars per year (at today's energy prices, it would be considerably more). All it took was a $3,000 sensor and a person to tag and catalog leaks. Over a year's time, we were able to reduce compressed air consumption to the point that two very large reciprocating compressors were placed on standby.

Ultrasonic detectors can also be used to test steam traps and sense electrical coronas and contact arcing.

Infrared Imaging

Another profoundly obvious fact is that heat is the enemy of machinery—particularly hot spots in electrical switch gear and transformers. Performing periodic thermal imaging will show you points that need attention before they become expensive failures. Again, comparisons from year to year are important to discover trends. Infrared imaging equipment can be very expensive and may well not be cost effective for a smaller facility to purchase. In those cases, the service can be contracted.

Motor Analysis

Analyzing the current waveform of an alternating current (AC) induction motor can yield information about the mechanical condition of the rotor and the electrical condition of the field windings. There are electronic analyzers available that use these data to diagnose potential motor problems before they become failures.

If your facility has a large number of AC induction motors, you should consider having them surveyed at least annually—especially the larger ones (100 hp and above).

Oil Analysis

Oil analysis is a reliable, inexpensive early-warning system. Trace metal contaminants in lube oil can point to component wear even before there are noticeable changes in vibration levels. This is particularly important for large machines with circulating lube oil systems. Once metal contaminants are detected, monitoring frequencies are increased to track the wear progression, and plans can be made to repair or replace accordingly.

The way the system works is you take samples from your lube oil reservoirs and send them off to a lab for analysis once a quarter or so. Again, it is important to keep track of developing trends.

PdM Summary

A comprehensive PdM system can result in

- Increased equipment reliability
- Longer equipment life
- Reduced costs
- Increased maintenance technician skills
- Reduced variability

Total Productive Maintenance

TPM can cover a wide spectrum of operator involvement in maintenance, from performing only the simplest tasks all the way to operators actually being the *primary* maintenance resources. In fact, if your operators were initially recruited and trained as maintenance technicians, learning to run the process equipment would be a snap. Then, when it is time to do maintenance, the people who will do the repairs know firsthand what the real operating problems have been. It is very easy to see how combining operator experience with in-depth technical skills would make for a potent workforce.

Having an operator workforce with a 100% technical background would seem to be highly unlikely for most of us. It is much more reasonable to assume that you have a traditional operating workforce who has only moderate technical skills. So, under these conditions, it makes more sense to implement the *operator-assisted* version of TPM. This implies that you still have a dedicated maintenance group, but they are assisted by operators who *they have trained* to make machine adjustments and do some of the routine repairs.

There are several benefits to this strategy:

1. It is backward compatible: You can implement this in an established facility simply by training *and certifying* selected operators.
2. It makes good operators better by providing more in-depth process knowledge.
3. It reduces on-shift delays and downtime.

4. It provides job enrichment and opportunity for operator advancement.
5. It permits having a reduced number of dedicated maintenance resources.

The following are some potential difficulties associated with operator-assisted maintenance:

1. It may result in labor relations problems with the maintenance workforce.
2. Union facilities will probably resist it.
3. It takes time and expense to do training and certification.
4. It requires extra pay for operator-technicians and has seniority implications since not all operators are capable.

In summary, TPM is highly desirable, but it will require some work on the part of management to implement properly. It must not be taken lightly as there are significant human relations implications that must be handled carefully to foster the kind of teamwork that is vital to make it a success. In any case, it is worth the effort because it offers huge benefits.

The Maintenance Process

Now that we have the strategies, we need a straightforward understanding of how they are put to use. Again, this is a highly simplified discussion and only intended to illustrate the flow of the maintenance process.

The basic maintenance process functions are

Requesting or initiating work
Estimating
Planning
Scheduling
Completion
Spare parts management

Work Orders

The work order is the instrument used to process a job from start to finish.

Work orders are used to

1. Plan and estimate work
2. Assign priorities
3. Capture maintenance cost and history
4. Document reliability issues
5. Schedule recurring PM tasks

Because it is so important to know where your maintenance dollars are spent, all maintenance work should be documented on a work order. Since anyone at a facility could potentially need maintenance services, it follows that anyone at a facility should be able to initiate a work request. Before today's technology, this was a paper system that could be burdensome, but that is a thing of the past. Today, submitting a work request can be as simple as sending an e-mail form to the maintenance planners. Now whether you actually do the work requested *is a management decision*. However, if the work order is declined, sound reasons must be given by management to the person who submitted the form—every time, period. Once a work order is accepted, it must be estimated and then presented to management again for final approval before proceeding to the planning stage.

Many work orders should be generated by the system even if your "system" is nothing more than a spreadsheet. In fact, 30–40% of the work done at any facility should be time-based PM. If this is not the case at your plant, you may have bigger problems than you realize. The ratio of man-hours expended on PM compared to all other maintenance work is a key benchmark.

Estimating

An essential part of the maintenance process is estimating the costs before the work is performed. Who would hire a contractor to make a home repair without first having an agreed to price? The same applies to plant work orders. It is typical for work orders to go through an approval process based on the estimated cost—higher costs requiring higher levels of approval. This is especially important for work that is not repair or preventive maintenance. For other types of work, like process improvement, management must be involved in the decision making and verify that there is a positive cost-to-benefit ratio. This assures that low-cost high-benefit jobs get the highest priorities and questionable pet projects get put on the back burner. Since management is ultimately responsible for bottom-line results, they will want to be participants in all matters with financial implications. While day-to-day routine work is the purview of the maintenance team, expensive modifications and rebuilds will necessitate management involvement. Rational decisions can't be made without knowing the costs involved.

Maintenance Planning

An old saying goes that you should plan your work and work your plan. Well, when you consider the cost per hour to retain a full-time maintenance workforce, this would seem to be very good advice. What is the definition of *planned work*? In the ideal, it means that a technician can arrive at a job at the appointed time, open his or her toolbox, and have everything needed to begin working immediately. If this were possible, it would be counted as 100% "wrench time" and would imply that the technician has the following:

- All needed parts and supplies
- The correct tools
- The required skills
- Equipment in a safe condition to work on
- Equipment that is sufficiently clean
- Sufficient number of people to assist in performing the task

If any of these are lacking, the work cannot proceed, and the 100% wrench time quickly evaporates. In addition, delays in completing the job could lengthen the amount of time the process is down, resulting in even greater expense.

In reality, it is impossible to have 100% wrench time, even for well-planned work. Realistically, the best you can expect is 50% to 60%. For unplanned work, you can expect only 15% to 20%, so it is easy to see why maintenance planning is necessary and beneficial—you get a lot more work done with the same number of resources, and there is less downtime.

Effective maintenance planning is dependent on the skill and experience of the person doing the planning. It is not one of those generic positions. If the planner has no experience in your facility as a technician, he or she will have a hard time putting together planned jobs that actually help people.

Even if your maintenance planning starts out a little rough, its quality and effectiveness can be improved over time if you can get feedback from the technicians each time a job is performed. This means that the technicians *have to write down what went wrong* on the work order and return it to the planner for corrective action. It then becomes imperative that every concern is corrected by the planner, so that over time, the job plan approaches

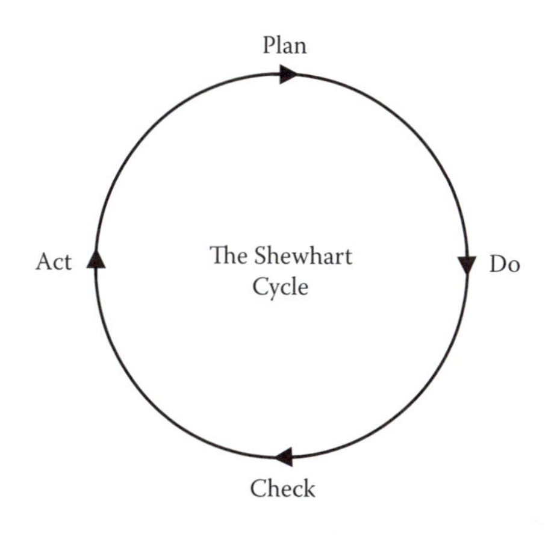

Figure 3.2 The Shewhart cycle.

perfection. This process is known as the Shewhart cycle and was frequently referenced by Dr. Deming.

The Shewhart cycle consists of four steps: Plan, Do, Check, Act (PDCA) (see Figure 3.2). You *Plan* your work (project, experiment, process change); *Do* the work (implement the change); *Check* (measure) the results; and *Act* on what needs to be changed. Repeating this PDCA process over time will result in a process or job plan that works.

This has significant implications. A long-time maintenance manager once told me that even in a large facility there are only 400 to 500 maintenance jobs that are repeated year after year. I have found this to be fairly accurate. Yes, from time to time you will encounter a new job, but if the facility is mature enough, those will be rare. Given enough time and diligence, it should be possible to develop preplanned jobs for 90% or more of all the maintenance work done on the process equipment. Think about how incredibly efficient your maintenance workforce could be if each one of those jobs were refined repeatedly by using the Shewhart cycle: sweet.

Work Order Scheduling and Priorities

Work order scheduling is the process used to assign *planned work* by calendar day—simple enough. However, on any given day, there are not enough maintenance technicians to go around, so it is also used for "resource leveling"—scheduling the amount of work each day to correspond to the number of people available. Scheduling is another one of those big topics that we only look at on a limited basis because of the need to stay focused on the "one thing": reducing variability. So, in that limited context, let us see how this relates to scheduling considerations.

Raise your hand if you do not know that work scheduling is done by priority. No hands raised? I did not think so. OK, so daily work schedules are done by urgency (what needs to be fixed first), equipment availability (production schedules), and resource (people) availability. The main point here is that *setting priorities is a team effort.* All functions—operations, maintenance, and technical—have to be involved. Every day is important in the improvement process, so it is necessary to start each day with a short meeting to agree on the day's objectives. Department barriers and territorial personalities are not allowed in this process (Deming's Point 9). Protecting the assets from damage is the highest priority. Second only to that is reducing variability. Third in line are PdM/PM chores.

With these ground rules established, the daily meeting is used to allocate maintenance resources accordingly. Maintenance planners should bring to the meeting a list of planned jobs that are at the ready to determine if any of them can be done that day. Also, running lists of issues to be addressed at future maintenance downtimes should be updated daily as new information is discovered and all of this incorporated into a longer-term master schedule. At the end of the meeting, assignments are made and small teams formed to address the day's issues. Any remaining maintenance resources can then be assigned to perform planned work from the schedule.

Remember, every day should be used to improve the reliability of the process. If your people are only able to "put out fires," your maintenance program is failing, and your situation is hopeless. The good news is that once you get beyond this, you will be able to schedule maintenance on a more rational basis—when it is due, not when you are forced to do it. And that is when real maintenance scheduling comes into play.

Work Order Completion: Getting the Feedback

There is one last issue that needs to be discussed before we leave this topic. It is easy to see that getting accurate, written feedback from the persons who actually did the work is vital information. It is an eyewitness account of the equipment condition "as found and as left" and how well the job plan went. It is also difficult to get because it seems that maintenance technicians love to fix things but hate to do the paperwork at the end of the job. Some even regard it with suspicion in the belief that it could be used against them in the future—another reason why driving out fear is so important (Deming's Point 8). Others simply regard it as drudgery. In any case, it is critical, and the entire organization should make capturing history a priority.

One way to measure work order feedback is for maintenance planners to scrutinize every returned job package for completeness and assign a numerical score like a schoolteacher correcting homework. It goes without saying that paperwork with missing information must be returned to technicians for a do-over. However, when this is necessary, the trail may be cold and the quality of the information degraded. To give this information the importance it deserves, completion scores for job packages returned *the first time* should become one of the benchmarks used to measure maintenance performance and reported to top management.

Inability to get accurate, complete work order feedback is a universal concern. At a symposium I attended a number of years ago, I heard one maintenance manager after another lament the fact that he could not get his technicians to close out their work orders properly. Finally, one person said that his facility had solved that problem, and it was a union shop. He said they valued the information so much that they paid a $1,000 bonus at the end of the year to each person *in a crew* that scored 90% on work order completions. If any one person in the crew did not make 90%, no one got the bonus. The kicker was that running completion scores for each person were posted on the department bulletin board.

Maintenance History

Maintenance history is simply the record of work performed on a piece of equipment. To be beneficial, it must include all of the events that have occurred over the operating lifetime. For each job that was performed, you need to know number of people it took to do the work; the date, time, and duration; the parts used; and the total cost. This last piece of information is vital. Accurate maintenance history enables you to calculate actual equipment operating costs and their contribution to the cost of manufacturing.

As history accumulates over time, you will begin to see patterns developing that indicate reliability status and whether operating costs are reasonable. Just like an old car, things wear out and become no longer "economically repairable." Without accurate maintenance history, you will not have the information needed to make rational decisions about whether to repair or replace.

Keeping all of this historical information accessible will be quite a chore if you try to do this with a paper system (not to mention filling up file cabinets). It is considerably easier to use a computerized maintenance management system (CMMS), and there are many available so we leave that for further discussion.

Spare Parts Management: Keeping Parts Visible

Having the *right* spare parts on hand is one the most difficult aspects of plant maintenance and is dependent on knowing the history of what has been used in the past. You can mine that history from a closed-loop work order system like the one discussed, and the result will be *accurate* parts lists for your equipment, which will accumulate automatically over time. Without accurate history, it will be extremely difficult to know what to stock and how many.

Knowing what to stock is only half of the parts management problem. Being able to locate a part in the middle of the night is the other half. The implication is that there are clearly cataloged storage locations set up, each part is assigned to one of them, and everyone who has a need to know can access the catalog. This seems simple, right? But, what if parts are withdrawn by technicians and not returned to stock? Technicians are notorious for accumulating their own personal store systems in their toolboxes and shop cabinets. This is because they want to be sure that they have a stash of the parts that they use frequently, and they do not trust a plant stores system to be their first line of defense.

Here is the problem: People can accumulate so much stuff that they forget what they have. If you doubt this, think about all the times you have bought new supplies for a home project only to discover weeks or months later that you already had them on a shelf in the garage. This happens every day in the plant environment on a much bigger scale. Once a person decides to keep personal private stock of some repair part, *it is no longer visible* to the rest of the plant population and may well even be forgotten by that person. In effect, it ceases to exist, and considerable expenses may be incurred in expediting a replacement for something you already have.

The cure is to make "all parts visible"—keep them in stores where they belong and not in people's toolboxes. All parts left over from a job must be returned expeditiously and placed back in inventory where they again become visible. If a shop needs to stock some frequently used parts and supplies, then it should have an official storage location that is managed by stores. The alternative is ever-increasing accumulation of parts and materials in shops and toolboxes, which is expensive and ineffective.

Stores Inventory Benchmarks

Unless you tightly control what is stocked in stores, inventories have a way of increasing exponentially over time. Inventory carrying costs being what they are, this is a big deal, so in many facilities adding an item to stores requires approval of the facility manager. The two questions routinely asked about whether to stock something is how often it will be used and what the consequences are for not having a spare. There are standard formulas for determining the former, while the latter is site specific and has to do with how long it will take to get a replacement in an emergency. If it takes 6 months or a year to get a replacement and your plant will be down until it arrives, you probably need to keep a spare for insurance.

Unless you measure and report the performance of your stores inventory, it will be difficult to know whether your plant stores inventory is effective. Here are a few of the common stores benchmarks:

1. Total inventory value
2. Number of inventory turns per year
3. Percentage slow-moving and excess items

Now, here is one that is less common but just as important, if not more so:

4. Percentage of stores items that are listed on plant equipment parts lists

Number 4 is hard to obtain without data mined with a CMMS. The main point is, if you have no formal parts lists (also referred to as bills of materials) for your plant equipment, how would you know whether you are stocking what is needed or even moving in the right direction? Ideally, about 70% of the items stocked should appear on one or more equipment parts lists. The importance of this cannot be overstated.

Computerized Maintenance Management System

From the preceding discussion on maintenance, it is easy to see that there are many data to manage. For all the years before computers, it was done manually—usually handwritten on big ledger cards. Work orders were handwritten forms with multiple carbon copies. The whole thing was a pain. These days, you can buy a CMMS for a relatively modest amount of money that will keep track of all of this for you. There are many of them out there with a wide range of prices. While they are all good, some are better than others. It is up to you to find the right combination of price and features.

Every CMMS out there will have the following as a minimum:

- Work order processing
- Work order planning
- Work order scheduling
- PM scheduling
- Plant equipment structure
- Plant equipment bills of materials with machine section locations
- Parts management
- Work order history

In my opinion, it just does not make sense not to have a CMMS. The one thing to be aware of is that it takes quite a bit of effort to set it up and keep it current. But, it is clearly worth the effort.

The Human Factor

Here are some lessons learned about the human side of maintenance management:

- The key to success is having the right first-line supervisors. It has often been said that supervisors have the toughest job in the organization because they have to represent management to their crew and represent their crew to management. It is a delicate balance. They must be people of the highest integrity and have consummate technical skills along with a fair amount of people skills.

  Because these people must lead their crews any time of the day or night, including weekends and holidays, it is critical that they are paid with the same overtime rules as the people they lead. Anything less is demoralizing and counterproductive. This is true for any first-line supervisor.
- Maintenance technicians should, in fact, be "highly skilled professionals." Ideally, they should be tech school graduates with superior verbal and math skills that will enable them to become experts in the use of today's advanced tools and test equipment. If your facility employs complex process equipment, provide appropriate training by the

manufacturers and create in-house experts who will be able to maintain the equipment in like-new condition and troubleshoot problems with confidence. This level of investment also shows employees that you believe in *them* and that you are counting on *them* to keep the plant running. It says that you think they are intelligent and valuable resources and not just "hired from the neck down." (I love that term. I learned it from a brilliant Flour-Daniel piping superintendent.)

- It is very important to keep maintenance crews busy with productive work every day. If your process is running well, that is the time to take PdM readings, rebuild spare components, and do construction projects. Maintenance planners must plan far in advance of the daily schedule and have stacks of planned jobs ready to go every week so there is plenty of meaningful work to do. Why is this necessary? This maintains good morale. This is how you ensure that the maintenance force can take pride in their work every day.
- You probably cannot afford to keep the number of full-time maintenance people on your payroll it takes to perform large-scale maintenance projects or new equipment installations.

  Here is the problem. Maintenance technicians love to do new installations because they provide the ultimate creative outlet. Skilled craftspeople finally get to "show their stuff" and actually build something. But, unfortunately, you cannot afford to let them be sidetracked from the main thing: reducing variability. While you can hire contractors to run miles of piping and conduit, they cannot help your operators make your process more reliable. For this reason, using plant maintenance technicians for new construction has to be limited to small-scale projects that are related to process improvement.
- To maximize the capability of your workforce, it is also highly desirable to be able to pool your maintenance resources for major maintenance downtimes in one department or another. This means that you have to eliminate any barriers that would prevent that. Does this sound familiar? See Deming's Point 9 again.
- Last and most important, maintenance is a *service* function. While this mindset must be instilled into the entire organization, it is even more critical for maintenance crews. Why is this the case? This is because we all know that there is a tendency for maintenance techs to become somewhat elitist. To combat that tendency, here are the key points I used to drill into my team. They made a huge impact.

1. We do not turn down work. The more work we perform, the more value we add.
2. The only acceptable answer to someone who is asking for assistance is, "When do you need that?"
3. If you as an individual cannot, for whatever reason, provide the assistance in the time needed, you must relay the request to a supervisor or manager, who in turn *must* get back to the requestor and his or her management and decide on the best course of action.

When my technicians realized that they were empowered to provide assistance, they stopped thinking of themselves as superiors and more as servants. In return, they *received* a lot more thank-yous—and they took more pride in their work.

Maintenance Performance Benchmarks

There have been all kinds of detailed, and what I feel are burdensome, maintenance benchmarks put forth over the years. Many of them have to do with measuring *how* the work is performed, as opposed to the actual *contribution* made by the maintenance function. In that vein, here are my picks for some meaningful benchmarks:

- Breakdown rate: The number and duration of breakdowns that occur each month. This is the one true measurement of maintenance effectiveness.
- Maintenance cost per unit of product produced: This is the *not* to be confused with total maintenance dollars spent, which is meaningless.
- Overall vibration levels: One number in inches per second that is the average of all vibration measurements taken in a month.
- The percentage of maintenance hours expended on PM and PdM: This tells you whether you are in the proactive or reactive mode.
- Work order completion score: One number that is the average of all completed work order scores in a month.

Putting It All Together

This has been a rather long discussion, so I want to summarize briefly what I think the perfect maintenance organization would look like.

1. The maintenance function is led by individuals with engineering degrees and broad industrial experience who understand their mission is to provide the highest levels of service and support the main thing.
2. First-line supervisors are mature, technically capable leaders who are trained and committed to the main thing and compensated using the same overtime rules as the people they supervise.
3. There are sufficient numbers of experienced maintenance planners to plan work far in advance so that technicians are fully employed each day.
4. Technicians are process experts who understand the main thing and are committed to providing the highest levels of service to the teams.
5. Morning meetings are held each day with operations to assign priorities, transfer information, and deal with issues.
6. Maintenance teams routinely work across department lines.
7. A contractor-partner is available to handle overflow maintenance work and construction projects.
8. There is a state-of-the-art PdM system in place.
9. There is a CMMS of sufficient capability to meet the needs of the facility.
10. Technicians fill in all appropriate information when closing out a completed work order and include feedback for improving the job the next time it is performed.
11. For each piece of plant equipment, there is a parts list or bill of materials of all the items needed for maintenance, and that information is tied back to the CMMS.
12. All on-site parts are in stores and available to all—not in private stashes.
13. There are insurance spares on hand for items that have long delivery times and would keep an operation down if they were not available.
14. Stores is managed to the standard industry benchmarks.

Chapter 4

Removing the Process Variability Barrier with Automatic Control Systems

Introduction

In every manufacturing process, there are variables that are continuously controlled in real time with hardware that is referred to as *instrumentation*, which is the generic term for the components within a "control loop": sensors, limit switches, transmitters, indicators, and controllers. An example control loop is shown in Figure 4.1, which is used in process and instrument diagrams (P&IDs) like the example shown in Figure 8.2.

Point 3 in our Model Vision states that our process is monitored and controlled with "modern instrumentation." One definition of *modern instrumentation* is a system that has control loops that can be interfaced to the site's computer network and supervised by a personal computer-based system like Wonderware®. In addition to improved control, this enables process variables and trends to be displayed on any workstation in the plant and stored in history. Then, management can see for themselves what kind of variability the plant is generating. To get an idea what this system might look like, see Figure 4.2.

Does this mean you have to update all the control systems in a plant just to get started on the main thing? Surprisingly, my answer would be no, but it is certainly desirable. In my opinion, you do not have to have the latest

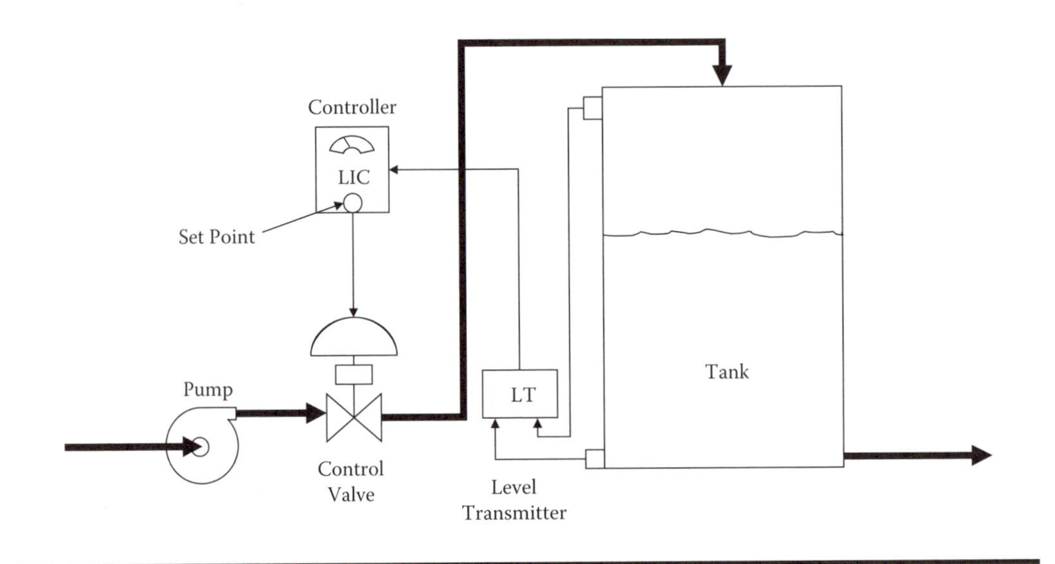

Figure 4.1 Instrumentation—components within a "control loop."

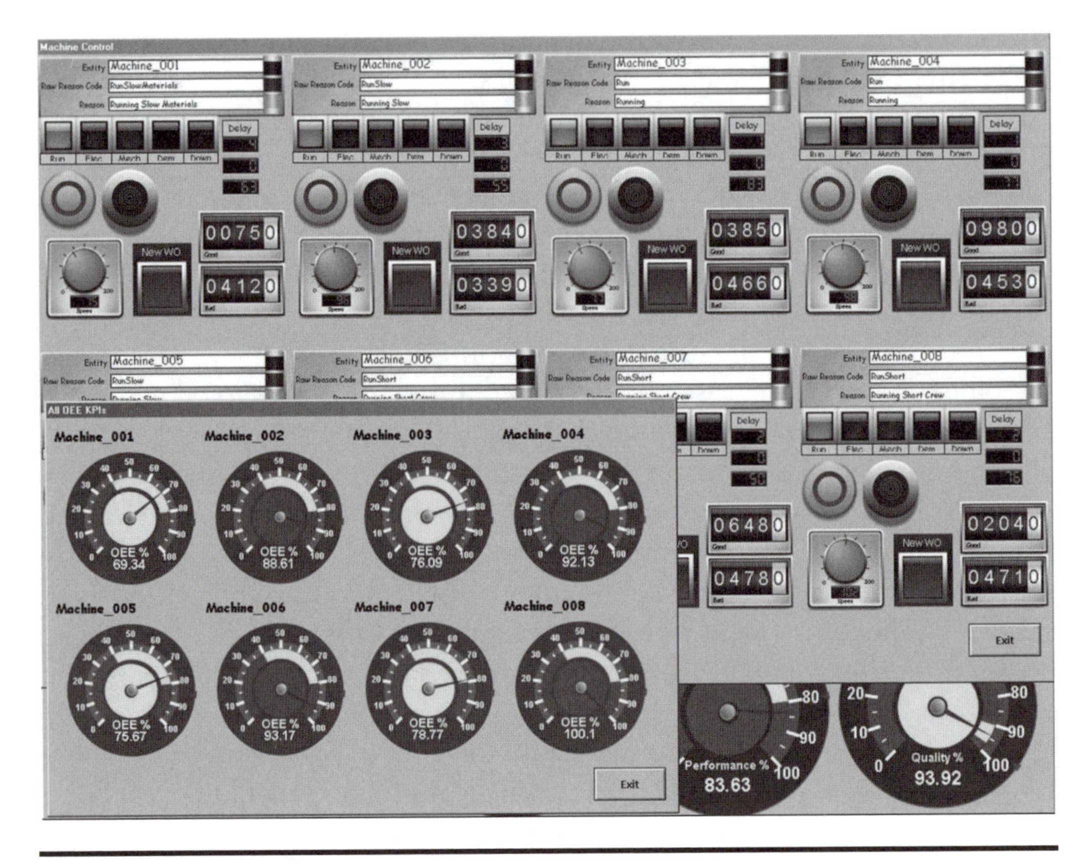

Figure 4.2 Process and instrument diagram.

electronic controls systems to have a stable system. Even quality pneumatic systems like Foxboro and Moore from the 1970s can work well if they are in good condition. However, maintaining them in that state may prove to be uneconomical and not worth all the effort. If you decide to take the plunge and upgrade everything, understand that, while electronic instrumentation has come down considerably in price over the years, the installation costs are still formidable.

Regardless of which path you take, the end point has to be same. A single operator must be able to sit at a desk and oversee everything going on in an operation with a personal computer, and the system must be visible on a network.

Getting Started

The process starts by you surveying every process variable you currently control with real-time closed-loop controllers. Some typical parameters are temperature, pressure, speed (or frequency), voltage, current, humidity, pH, and maybe concentration. You will need to have a team verify, by monitoring with data-logging equipment, that each and every variable is stable—free of oscillation and able to track step changes accurately. If your employees are not able to do this on their own, then you will need to contract to have it done. Spend whatever time and money it takes to do this survey accurately because everything following depends on it.

If during the survey any process variables are found to be erratic, here are some potential causes:

1. Uncalibrated or deteriorating sensors
2. Faulty controllers and interfaces
3. Improper control tuning
4. Excessive play or looseness in final control elements (valve positioners and servos)
5. Loose or corroded wiring connections
6. Leaking actuator cylinders and diaphragms or connecting tubing
7. Improperly sized control components (either too big or too small)
8. Improperly sized "prime movers" (motors, pumps, actuators)

Let us look briefly at each one:

1. Many sensors wear out, corrode, or just deteriorate over time. Unless you have a systematic calibration program that checks them all at least once a year, you may be missing something that is causing huge problems but is going undetected. Knowing that your sensors are telling the truth is the starting point for everything else.
2. Occasionally, controllers or their interface modules can fail or partially fail and become erratic. About the only way to verify this is to replace one with a spare and see if things improve. While this is somewhat rare, it does happen, and for that reason you will need enough spare control components on hand to be able do the swaps.
3. Control loop tuning is a bit of an art form. It also requires specific knowledge in the field of instrumentation and a feel for how the tuning parameters interrelate to each other and the rest of the process. There are a number of theories on the best procedure for determining the optimal settings, which is best left for those that are trained in this subject. It is sufficient to say that you will need resources on your team who are trained in loop tuning and have the skills to be effective and not do more harm than good.

   Also, be aware that there are people in any organization who like to experiment with controller tuning after hours when they have some spare time on their hands. Do not be surprised if unsecured controller settings mysteriously change.
4. Even after everything is calibrated and properly sized in a control loop, if there is play in any of the linkages in the final control element, the loop and the rest of the process will still oscillate. The fix, of course, is to tighten or replace worn or defective parts. The standard loop calibration process should find almost all of these as a matter of routine. Just be aware that this is one that is easily overlooked.
5. Loose wiring will generally cause intermittent control problems that are always the hardest to diagnose. When a controlled variable exhibits erratic behavior even after calibration, it will be necessary to check each of its wiring components point to point using a signal generator and oscilloscope. This is tedious but it is your last resort.
6. Leaking cylinders and diaphragms and any other air leak in a noisy plant can be found with an ultrasonic tester, which was discussed in detail in Chapter 3 in the section on predictive maintenance. Hydraulic leaks, on the other hand, are quite visible. Any leak is a problem and needs to be fixed.

7. Each element in a control loop must be sized to operate about a midpoint. If a control valve is too big, it will have poor turndown and behave erratically at low flows. This is also true of flow measurement sensors. If you pair both of these together in the same control loop, you could end up with a very unstable process at low rates of operation. On the other hand, components that are too small simply cannot "keep up" at higher rates. A valve cannot pass enough volume, a servo cannot react fast enough (insufficient "slew rate"), and so on.
8. Improperly sized prime movers like motors and pumps can result in the same type of erratic behavior as improperly sized control components. The main difference is that if they are too small, they are probably overloaded and will eventually fail in the middle of the night during a production run for the biggest order you have ever had.

A Few Real-Life Examples

Automatic control opportunities exist everywhere, and at the various facilities where I have worked over the years, there are a few that stand out that are worth sharing:

1. At a large petrochemical plant, there was a central, outdoor lube oil system for several very large, steam turbine-driven air compressors. The problem was erratic temperature control in the winter.

   Lube oil systems must be maintained at constant temperature to keep the oil within specified viscosity ranges. In addition, the oil must be kept warm enough to preclude water vapor from condensing and accumulating in the bottom of the reservoir. The plant was located in a very hot climate, so the cooling water control valve was sized to provide sufficient capacity to keep the oil cool during summer conditions. However, during winter conditions, the valve was too big. The slightest valve opening caused an inrush of very cold cooling water, and since the oil system was located outdoors, very little cooling water was needed. The temperature chart looked like the proverbial square wave.

   The fix was a "split range" controller. A smaller water control valve was piped in parallel to the large one. This new valve was sized for winter flow rates. The controller operating range was divided between the two valves proportionate to their capacities. During low-demand periods, the system operated on the smaller valve. As cooling water demand increased, the system seamlessly opened the larger valve to augment the smaller one. After that, lube oil temperatures were well controlled regardless of ambient conditions—problem solved.

2. At a polymer chemical plant, a 10,000-gallon reactor was heated with Dowtherm® synthetic oil. The oil was heated to 500°F in a gas-fired unit that ran at maximum firing rate during the initial reactor heat-up period. However, once a certain temperature was reached, the chemical reaction initiated, and the process became highly exothermic. At that point, the Dowtherm heater would go to minimum firing rates for several hours and then transition to a low, steady heat rate until polymerization was completed. During these times, the gas pressure regulator valve had insufficient turndown, and gas supply pressures became erratic. Often, the flame became unstable or would simply go out, tripping out the heater, which would have to be manually restarted. This resulted in considerable variability in the reaction and polymerization process and affected finished product consistency.

   The fix was another split range control system. A small gas pressure-regulating valve was piped in parallel to the large main valve. The internal spring ranges in the two regulators were adjusted to divide the operating pressure ranges. The large regulator was set to begin opening as soon as there was a slight pressure drop, indicating that the small valve could no longer keep up. As soon as the chemical reaction entered the exothermic period, the heater went to minimum firing rates and operated on the smaller valve. During the polymerization phase, constant gas pressure was easily supplied by the smaller valve, enabling smooth, accurate reactor temperature control. Product variability was greatly reduced—problem solved.

3. At that the same chemical plant, each of the reactors was fitted with packed distillation columns to recover ingredients that flashed off during the boiling exothermic period. The problem was "column flooding." (A column is flooded when it fills up with liquid to the point that there is severely reduced free area for incondensable gases to pass through. This flow restriction causes the differential pressure across the column to spike and can result in severe internal damage to the fragile column packing.)

   A water-cooled condenser was bolted to the top of the packed column. The intent was for incondensable gases and lighter reaction by-products to pass through the column and out of the system while recovered reaction components would condense at the top, fall back onto the column packing as "reflux," and eventually drain back into the reactor.

Performance of the packed column was heavily dependent on establishing and maintaining an equilibrium point so that the packing rings were wetted for maximum surface area and contact with the rising condensable components. This was accomplished by controlling the cooling water to the condenser to maintain constant temperature at the top of the column. However, during the exothermic period, when a large volume of reaction components were flashing off, the condenser had sufficient cooling capacity to severely flood the column. On more than one occasion, the column internals were destroyed.

After careful review, it was decided to place a limit on the amount of cooling water that could be applied to the condenser regardless of how high the temperature went at the top. This was accomplished through a control strategy known as a "low selector switch." In this setup, two controllers were used to control one final control element (in this case, the condenser cooling water valve), and the switch chose the controller with the lowest output. Through experimentation, the optimum maximum water flow was established, and it became the overriding limit. The result was the column could not flood, while reaction times were reduced along with greatly reduced product variability—problem solved.

4. At a paper mill, a high-speed paper machine was equipped with a headbox design that contained a large drilled block of stainless steel as a flow distributor for the forming section. The problem was that, due to the characteristics of this design, the jet-to-forming fabric relative velocity varied over time, affecting the characteristics of the paper. Variability was quite high, and defects were common.

   A large centrifugal "fan pump" supplied a 20,000-GPM dilute mixture of pulp and water to the headbox, which was injected into the forming section in a thin, high-speed stream referred to as the "jet." This situation occurred 30 years ago, and at that time, it was standard practice to control fan pump speeds as a ratio of overall machine speed. Operators adjusted the fan pump speed ratio up or down based on measurement of paper samples taken periodically. However, what the jet-to-fabric velocity actually was at any point in time was based on past measurements and amounted to an educated guess. In any case, variability under this control scheme was much higher than desired.

   As an experiment, it was decided to install a pressure transducer on the headbox just after the diffuser block and at the point of lowest velocity. Theoretically, the static pressure at that point would be closely proportional to the actual jet velocity. That pressure signal was then used as the feedback to control fan pump speed to maintain constant pressure. Using some simple math, the pressure/velocity characteristic curve was then programmed into the machine computer so operators could continue to adjust jet-to fabric velocities as before. As a result, strength variability was cut in half. Quality was increased, and defects were all but eliminated—problem solved.

5. The paper machine discussed was supplied fiber by a parallel plate "refiner." (A refiner brushes and cuts the pulp to size to increase the strength of the paper). The problem was inconsistent refining levels, high paper strength variability, and defects.

   The amount of work done on the fiber, or the amount it is refined, is inversely proportional to clearances between the grooved refiner plates. In other words, the closer the plates are, the higher the refining level and correspondingly higher paper strengths will be (up to a point). Now, the plates in this particular refiner happened to be hydraulically loaded, so that the incoming pulp stream pressure had a direct effect on plate spacing. Higher pressures forced the plates apart and resulted in less refining, while lower pressures caused decreased plate spacing and more refining. (This problem was eliminated in later refiners through mechanical plate loading.) But, there was more.

   A *manual* valve at the refiner inlet regulated a stream of recycled fiber back to the supply tank (or "chest"). Operators set the refiner inlet pressure by opening and closing that valve. The problem was that, over time, fiber accumulated in the internals of the valve, restricting the flow and causing an increase in pulp supply pressure to the refiner. Operators had to keep an eye on the pressure and make frequent adjustments to avoid completely unloading the refiner and making 100% defective paper. In spite of their best efforts, they could not keep the refiner inlet pressure constant.

   What was the fix for this nagging, years-old problem? Some old parts lying around in the instrument and electrical shop were the solution. With only one Foxboro Model 43A field pressure controller, a diaphragm pressure transmitter, and a spare pneumatically operated "pinch" valve, refiner inlet pressure was automatically and forevermore controlled at a constant, and strength variability was reduced by two thirds—problem solved.

Conclusion

You cannot even *begin* to make real progress in reducing variability unless all of the process variables like temperature, pressure, and the like are flat-lined. To do otherwise would epitomize the proverbial exercise in futility. So, one way or another, you must gain complete confidence in your plant control systems and *know* that all process variables are stable 24-7. The easiest way to do this is to upgrade all of your control hardware with the latest electronics. If you cannot do that because of cost constraints, you can achieve similar results, but you will have to work a lot harder on maintenance. In either case, the personal computer-based supervision and network visibility are *not* optional. So, if you cannot interface to your current system, you are stuck with replacing it.

Chapter 5

Removing the Product Variability Barrier with Statistical Process Controls

Introduction

While the previous chapter on real-time process control was relatively straightforward, this section is relatively complex, tedious, and never ending. In fact, getting all of the known hardware problems fixed is merely a prerequisite for the main event: improving the quality of your product. This is where statistical process control (SPC) takes over. SPC enables you to dig much deeper to find the root causes of product variability and is therefore an essential element of modern manufacturing.

I must preface this section on SPC with an emphatic disclaimer. I am *not* a statistician—not even an *apprentice* statistician. I will only say that I have been exposed to statistical methods and have seen them work marvels, so you could consider this to be more of a testimonial than a treatise. I hope my hands-on experiences will enable me to show you some of the tools and how you might go about applying them. However, when we are through, if you decide to try some of it in your plant, you will need considerably more knowledge than what is presented here. Dr. Deming provided a wealth of that knowledge, there for the taking, in *Quality, Productivity, and Competitive Position* (*QP&CP*) (MIT Center for Advanced Engineering Study, 1982). And, of course, there are *many* other fine resources out there, including Appendix 2 from Statit®. Just do not spend all your time reading books instead of rolling up your sleeves and getting to work. Eventually, you may find you need the services of a professional outside your organization.

Sampled Data versus Continuous Monitoring

To begin, let us establish a fundamental concept that every product or work in process, whether a physical object or a service, has characteristics that define its *value* to the customer. These are generally referred to as *product attributes*. And, while there may be some exceptions, generally these attributes cannot be measured on a continuous basis. They can only be sampled and measured as shown in Figure 5.1.

How you perform the sampling and what you do with the results will determine your success or failure as an enterprise.

In the previous chapter, process variables were measured continuously in real time and with an accuracy as high as 1%. If we displayed those variables on strip chart recorders, each variable would be connected to a colored pen that would begin tracing its value as a continuous timeline. At any point in time, you could see the pens move almost instantaneously in response to changes in the process. If a variable is constant, its pen moves little and traces a straight line. On the other hand, if a variable is noisy, the corresponding pen will oscillate rapidly, making a much wider line. (This is just for illustration. Remember, we supposedly fixed all of those in the previous chapter.)

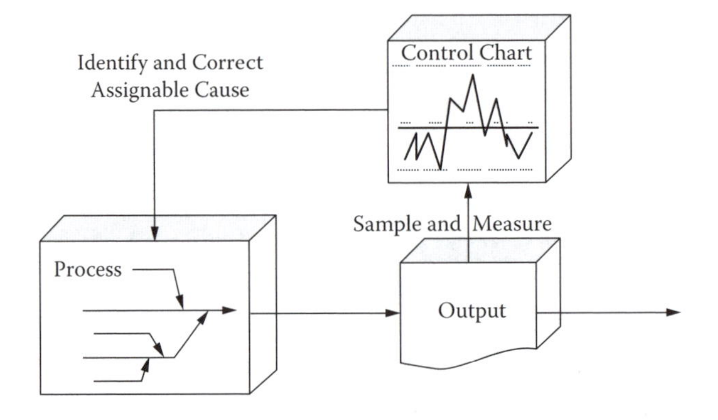

Figure 5.1 The control chart process. (Courtesy of the Data and Analysis Center for Software [DACS] Gold Processes, http://goldpractice.thedacs.com/practices/spc/index.php.)

Now, think of the stream of products or parts coming off your production lines. Each item is defined by a set of characteristics or product variables (size, weight, color, texture, fit, finish, etc.), very few of which can be measured in real time. The rest must be sampled and compared against a standard by a human; therefore, the noise in these data streams is orders of magnitude greater than the machine-controlled process variables. In fact, from a "metaphysical" standpoint, there is no single, definable value at any point in time. You can get close to one, but that is about it. Each measurement (or inspection) is influenced by the person doing the measuring, the repeatability of the test, and the accuracy of the standards, to name but a few factors.

Thinking back to the strip chart analogy, let us now visualize the product characteristic data. Instead of continuous lines, they are streams of noisy discrete points—samples—that form wide bands. Each band has a center defined by a statistical mean and a width approximately equal to plus or minus three standard deviations: sigmas. Within this bandwidth reside 99.7% of all possible values, which takes into account every influencing factor—those we can see and those we cannot.

Again, *your task is to drive variability lower and lower*, which will evidenced by correspondingly lower sigmas. If you are successful, eventually, when your people sample your products, the values they measure will always be within specifications. They will not be able to find any defects because those values will lie outside the range of plus or minus three sigmas for your process. They will not exist in your universe *because they cannot.*

The Tools

Nine commonly used SPC tools are listed in this section that we discuss in this chapter. In addition, Statit Software, Incorporated, has developed an excellent SPC overview paper that provides additional details and can be downloaded from the Internet. For convenience, a copy is provided (with their kind permission) as Appendix 2, and I refer to it throughout this section.

There is a logical order in which the tools are applied, and they are listed pretty much in that order:

1. X-bar and range chart
2. Histogram
3. Pareto chart

4. Control chart
5. Fishbone diagram
6. Flow chart
7. Process capability study
8. Designed experiment
9. Scatter diagram

Next, I try to lay out a straightforward path for using these tools to implement SPC from start to finish. As before, the intent is to show you how these can improve your process and become part of your manufacturing strategy. To implement this properly, you will need the services of an SPC professional.

Prework

The process begins by you taking another survey, only this time it is of all of the known product variables—dimensions and attributes—that must be kept within specified limits. Each operation in your manufacturing process must be treated separately, and you will need plant or company *experts* assigned to each one. You will know that your surveys are complete if you can answer "yes" to each of the following questions:

1. If we controlled this set of dimensions or specifications perfectly, every time, would this *positively guarantee* that the parts would fit at the next assembly point or flow smoothly through the next processing step?
2. Or, if these variables were controlled to specification, would our *finished product* meet specification?
3. If these specifications were met every time, would our finished product have attributes that our customers would *recognize* and *appreciate*?

When your surveys are completed, you should have before you separate lists of all variables associated with each product or process step. We will assume the combined stack is formidable, and it is impractical to tackle it all at once. So, now you will have to decide where to get the biggest bang for the buck.

Look ahead by having your team rank all of the variables by known problems: highest percentage defective, highest customer rejects, or even known attribute failures. Specifically, which dimensions or specifications have been difficult to control? Do not guess. Scour your records to be sure.

If you do not already have SPC software (like Statit) installed on your plant computers, now is the time. Do the homework, make an intelligent selection, and get this set up before proceeding. Incidentally, you will need personal computers (PCs) set up at every point in your operation where you intend to collect and chart data. Yes, you could resort to paper charting and statistical calculators, but the cost of the paper charts and the time to manually record the data are not worth it. PCs are just too cheap these days, and you will get much more accurate results.

A Word of Caution about Sampling

Getting valid statistical information about your process is dependent on obtaining representative samples at regular intervals. Yes, sample sizes must be statistically significant, but just as important, individual samples must be randomly selected. In fact, the best sampling is done automatically by the process. While this is ideal, it could take a while to modify your equipment. So, to get your SPC program up and running right away, you will probably be sampling manually.

All too often, employees pick samples with a bias—one way or the other—those that look the best or those that are clearly defective. It is critical that you train your people how to sample properly and give them a written procedure to follow in a process control manual (which is covered in detail in Chapter 8). *Initially, you will need to audit the sampling process frequently to be absolutely sure everyone is doing it consistently and correctly.*

Getting Started

Step 1: Set up entries in your SPC software organized by process step for each variable found in the surveys.

Step 2: Establish a sampling protocol. To do this, you will need to establish the following:

- Correct sample size: Usually, this is the minimum size that is "statistically significant," typically three to five items for discrete samples.
- Sampling interval: This depends on your capability to process samples but is typically once per hour.
- Sampling protocol: This involves how samples are selected to minimize bias.

Step 3: Start sampling and entering the results in the SPC program.

SPC Tool 1: X-Bar and R Charts

When the data start accumulating, the program will begin generating the various statistical charts and tools, the first of which are X-bar and range charts.

X-bar and R charts are actually two charts plotted on one page. An example is shown on in Appendix 2. On the X-bar chart, each point represents the average of a group of individual samples taken at a single point in time. The bottom graph is the corresponding range, which is the difference between the highest and lowest values in a sample group and is indicative of relative variability. By the way, if your sample size is large (10 or more), you will want to use the X-bar and sigma charts instead of X-bar and R charts.

For each shift and day's data, you must calculate the *cumulative* mean (X double bar) and standard deviation (sigma double bar) for all of the samples taken. After a few days, you should begin to get a feel for where the high-variability opportunities are.

Here are some possible outcomes:

1. The standard deviations and corresponding calculated percentage defectives are consistently high.
2. The cumulative average values are not at the desired target; they are either consistently higher or consistently lower.
3. The charts are different from one crew or shift to another.
4. The process runs fine for periods of time and then experiences upsets.

Each of these has a unique course of corrective action.

If sigma is high, it goes without saying that variability is high, and the best way to see this is with a histogram (SPC Tool 2). The main question is, does the span of six sigmas exceed and envelop the allowable range of values for your specifications? If it does, then you are ensured that the system is producing defects. You may have difficulty finding them, but they are there, and some of your customers know about them firsthand. The excessive variability could be due to shortcomings in process capability or overadjustment by the operators. If it is the latter, use of a control chart (SPC Tool 4) might just resolve it.

If the cumulative average values are consistently higher or lower, the process may simply not be capable of running at the specified target. A capability study (SPC Tool 7) or designed experiment (SPC Tool 8) will determine the answer. If the study determines that the process is, in fact, already capable, then you have an operator training problem, which is a lot cheaper to fix.

If the charts are different from one shift to another, in all likelihood you have a training problem. You will need to hold a group meeting that includes members from all teams so they can compare notes and find out what the differences are. After that, document your findings and train each group to use the same procedure.

If the process runs fine for periods but has upsets, do the upsets occur on all shifts? If they do, then it could be process related. If it happens only on one shift, you will need to assign someone to observe and investigate. Again, it could be a training issue.

X-bar and sigma data are the heartbeat of your manufacturing process. Everyone in the organization needs to understand this, and you will need to accumulate and report these two values daily for each shift, for each day, week, and month. Do not forget Deming's Point 8: You must drive out fear if you expect to get honest reporting and make progress.

For reporting purposes, X bar must be expressed as ratio to target. The value 1.0 means the variable is at target. Values above target are a number greater than 1; values below are less than 1.

The total average cumulative value of X double bar for all monitored variables along with the total average sigma (sigma double bar) must be charted and made a part of the monthly report to top management. *Top management will need to be trained on their importance so that they can provide the resources necessary to drive them to increasingly more favorable values.*

SPC Tool 2: Histogram

Another chart readily available from your SPC software is the histogram, which shows the "shape" of the data distribution (see Appendix 2 for an example). It allows you to see almost at a glance whether the data are normally distributed, where most of the values lie, and what values you can expect for upper and lower limits. Data that are skewed or asymmetric may point to a particular cause that can be corrected relatively easily once it is known. See Appendix 2 for more details.

SPC Tool 3: Pareto Diagram

As you accumulate more X-bar and Sigma data over time, you will discover where your biggest opportunity areas are. Since you cannot work on everything at once, you will need to rank them by severity. Here are some possibilities:

1. Highest known reject rate
2. Highest known customer complaints
3. Highest calculated percentage defective
4. Highest deviation from target
5. Highest sigma

Now, with some reliable data at your fingertips, you can chart each category on a Pareto diagram for a visual aid and publish your strategy for attacking the worst problem areas. All will know you have a plan for improving the situation and can rally behind it (Deming's Point 10). See Appendix 2 for an example.

SPC Tool 4: Control Charts

Previously, I stated that overadjustment can increase variability. The cure for that particular problem is control charts. See Appendix 2 for examples of what they look like. A control chart is very similar to an X-bar chart except it has two additional lines, referred to as upper and lower control limits. The traditional placement of the control limits is three sigmas, which is not too surprising. (However, for low-variability processes, this interval is sometimes tightened.) The value used for sigma in this case is the long-term, *cumulative* standard deviation. A process is considered to be in "statistical control" if it stays within the control limits when left to itself without adjustment or any other intervention. By the way, you cannot actually *do* control charting if your process *is not already in statistical control.*

The main purpose of the control chart is to show an operator when it is *appropriate* to make an adjustment to avoid overcontrol. Overcontrol occurs because all process adjustments are performed "after the fact." There is no way to know what is taking place in real time if the only feedback information you have is based on a group of samples taken 15 or more minutes ago. Consequently, an adjustment made after a single sample may well drive the process in exactly the opposite direction of what is desired, reinforcing rather than counteracting the process swing. The true state of the process has to be confirmed by a series of sample points that establishes a trend or a shift in centerline. The exception to this is when one or more points lie beyond the control limits. Then, the data indicate a condition that lies outside the established plus or minus three-sigma envelope and warrants immediate corrective action.

If implementing a control chart results in a higher average sigma for a process variable or product attribute, the control limits are set incorrectly, the control strategy is erroneous, or the operators need more training.

SPC Tool 5: Fishbone Diagrams

Solving complex problems like the ones you will be dealing with requires a disciplined approach. Again, volumes have been written on this subject, and everyone has a favorite tool. However, the one I personally prefer is the fishbone diagram because it is the easiest to use and, in my opinion, the most straightforward.

The fishbone diagram (or Ishikawa diagram) displays the causes and effects and interrelationships of the components of a process. See Figure 5.2 and Appendix 2 for an example of one in the early stages of development. They are an excellent vehicle for pooling the collective knowledge of a team of workers who are *intimately familiar* with a process and trying to determine the root cause and effects. Done properly with a trained facilitator, team members should be able to build on each other's inputs so that the summed knowledge is greater than the individual parts.

The diagram starts with a big horizontal arrow pointing to the right to a label that represents a problem to be solved. Slanted vertical lines are

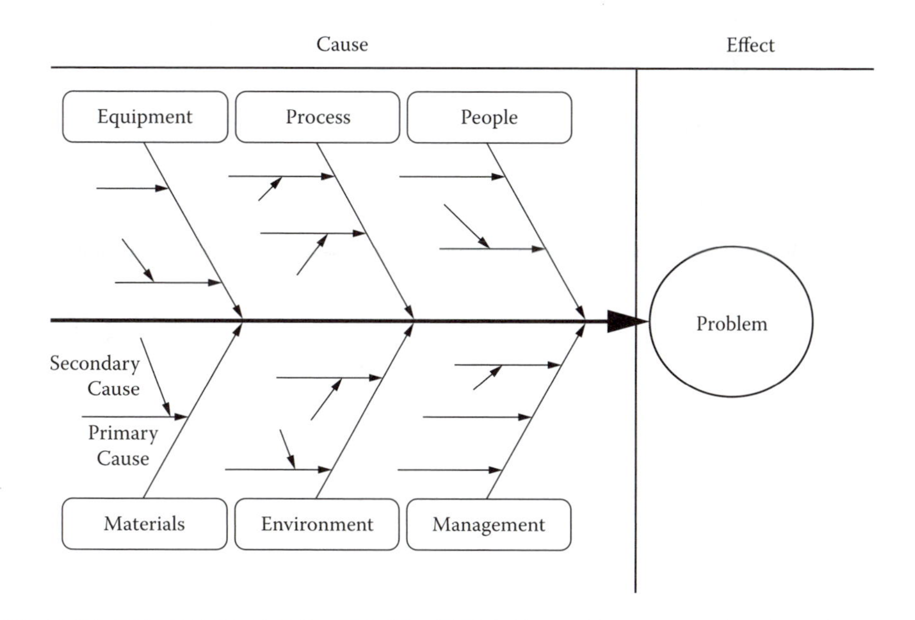

Figure 5.2 Fishbone diagram. (Courtesy of Wikimedia Commons.)

added in a herringbone pattern that represents what the team thinks are the main factors that influence the problem. One by one, each factor is explored and details added as new branches that probe deeper and deeper into the process until the knowledge of the group is exhausted or all pieces of the process have been accounted for. The group then prioritizes the potential causes regarding which ones are the most likely to have the major influence.

Plans of action are then formulated for activities by small "committees" of team members *who have the most in-depth knowledge* of the specific area in question. Their job is to investigate the premise of each proposed cause and effect and report back to the main group. Over time, each branch is explored and either validated or ruled out as actually having an influence. Those that have been ruled out are erased so that eventually only verified interactions remain. *The only true verification method is to make an adjustment and measure the system's output response.*

Each verified cause-and-effect relationship must be further investigated by *technical resources* using more of the SPC tools, such as process capability studies, designed experiments, and scatter diagrams to try to *quantify* the actual system responses. The goal is to be able to intelligently reengineer processing equipment, raw materials, or both until that particular source of variability has been "beaten into submission."

SPC Tool 6: Flow Chart

A prerequisite for actually being able to put together a fishbone diagram is, of course, a clear understanding of how the process *actually* works as compared to the often-erroneous version of how it is *supposed* to work. If you have any doubts about this, maybe you had better develop a process flow chart (SPC Tool 6) like the one shown in Appendix 2 just to be sure. One of the things it incorporates is logic decision points that may provide added information to a problem-solving team.

The X-Bar and Sigma Charting Marches On

As variability sources are discovered and resolved, X-bar and sigma values should trend more and more favorably. If they do not, then the real sources have either not been identified or the corrective actions taken were ineffective. But for now, let us assume that we are winning the war. As soon as you know that you are making progress, share the good news with the teams and the general plant population. Celebrate. Congratulate. Most of all, *be appreciative.* Revise your top three priorities as appropriate and continue.

Advanced SPC Tools for Digging Deeper

SPC Tool 7: Process Capability Study

Process capability studies are used to determine whether your process is robust enough to reliably produce products day in and day out that are defect free. As stated in the introductory comments, if your process control is tight enough, then defects cannot exist statistically. Appendix 2 shows how Statit software does the data crunching to do this for you. Yes, there are other methods, but this is one example. To obtain representative capability data, you must first be assured that your process is running smoothly in statistical control and is free from extraneous upsets and special causes.

SPC Tool 8: Designed Experiments

Designed experiments are primarily used to find an operating "sweet spot." In other words, is there a particular operating point that gives the highest output, the lowest variability, the highest quality, or even the lowest energy consumption? Clearly, it would be beneficial to know where these points exist. How about being able to correlate a product *attribute* (which is what the customer is looking for) directly to a *process variable* that you monitor continuously in real time? Once the relationships are established, you can know in the future that the attribute will be consistent as long as the corresponding process variable is controlled at the values found in a designed experiment.

You run designed experiments by first establishing the process component interrelationships using fishbone diagrams so that you can *intelligently* theorize what *might* be possible if certain settings were fine-tuned—hence the term *designed.* Then, you test your theory by varying one process setting, holding the rest as constant as possible, and measuring the system response. If you are ambitious, you can even try varying two settings, but it gets to be a bit much beyond that. In any case, designed experiments are reserved for stable processes that have already achieved statistical control and are not in a state of transition. Also, this is where the help of a trained professional may be beneficial to obtain reliable results.

SPC Tool 9: Scatter Diagram

Scatter diagrams were alluded to in the section on fishbone diagrams. They are used to establish specific relationships between two process variables and may be useful for finding an operating "window" where variables have a little room to move around without causing major upsets. This is in contrast to a process that is dependent on finding a single point where all the planets have to be aligned to make good product. The diagram is created by holding all process variables as constant as possible, varying one of them and measuring the response of one or more others. The test should be repeated more than once to confirm the relationship. The points are plotted on a chart, and using curve-fitting software, an attempt is made to draw a line that best approximates the relationship.

That is it for the tools. Again, it is an overview. Each of these tools requires some study and practice to implement with confidence. And, because statistics is a complex, math-based field of study, you will need a professional to guide your organization through the process. It is worth the effort because without solid SPC, you will never be able to reliably control product variability.

Summary

The purpose of this chapter was to provide a high-level road map for attaining statistically controlled manufacturing processes. The steps are

1. Provide networked PC workstations throughout the manufacturing process.
2. Install SPC software.
3. Implement a massive employee training program on statistical methods.
4. Create variable and attribute lists for each manufacturing step.
5. Review product defect history for known quality problems.
6. Load variables and product attributes into an SPC software database.
7. Establish well-defined sampling protocols and verify with periodic audits.
8. Begin computer charting all essential controlled variables and attributes.
9. Begin publishing summary data for deviation from target (X double bar) and standard deviation (sigma double bar).
10. Use items 5 and 8 to determine the top three problems in each manufacturing area.

11. Develop management's plan for reducing variability and high-priority improvement items and publish for employees.
12. Establish control chart limits as soon as sufficient data have been accumulated.
13. Decide on control chart control strategy, train operators thoroughly, and begin control charting. Audit the process, retrain periodically, and make adjustments as necessary.
14. Assign technical resources to focus on the top three problems using SPC tools. Obtain outside assistance if needed.
15. Update the top three problem lists if priorities change as a result of improvements achieved by control charting.
16. Begin the Plan, Do, Check, Act (PDCA) improvement process for the top three problems in each operating area.
17. Measure and report progress to top management and employees.
18. Celebrate successes whenever possible.
19. Continue the improvement process daily.

All employees, especially top management, must be involved in the improvement process. If this happens, reducing variability will become as important to the organization as productivity, safety, and customer service. If not, the process will fail and become another "program-of-the-month" exercise. It has been proven time and again that reducing variability is very good for business and bottom-line profits because quality processes produce quality products at the lowest costs. Choosing not to embrace this has long-term negative consequences. If you have doubts, Google GM and Chrysler and see if you can connect the dots.

Chapter 6

Removing the Raw Material Quality Barrier

Introduction

Next to process equipment and technology, there is no other aspect of manufacturing that has a bigger impact on the quality of finished product than raw materials. Again, this concept would seem to be patently obvious to most of us, and yet many raw material purchases are still based on price instead of quality. But, the fact is you have to have both to survive. It does no good to have the finest raw materials on the planet if the costs are so high that your product cannot be sold at a profit. There has to be a balance based on *partnership* between you and your supplier.

Dr. Deming says that a supplier worthy of being your business partner is capable of supplying materials that consistently meet specifications. And, the supplier can prove it by providing the statistical process control data (just like yours) for each batch. In addition, your supplier will give you the very best price possible to ensure that you are *competitive* and will be able to *stay in business*. The supplier does not do this to be altruistic but because it is how *the supplier* plans to stay in business. In return, the supplier will not have to worry about you continuing to look for a cheaper supplier. The supplier will not have to wine you and dine you and *hope* that it will have your business next year or next month. You trust each other because you are partners who are in it for the long haul. If this all sounds like a fairy tale, I can assure you it is not. This is how the Japanese (who were trained by Dr. Deming) operate and is one of the reasons they have been able to maintain their high quality levels year after year. I have seen it work firsthand.

Now, let us look at the flip side and say we are only going to buy from the suppliers that give us the lowest price. In your personal life experience, has this *ever* worked? Have you *ever* purchased *anything* that was "cheap" that did not give you buyer's remorse? What you can expect from the low-cost bidder is high variability and nonconformance to specifications: defects. Why? Because this bidder cannot afford to invest in quality equipment and the best technical people and has no interest in statistical process controls. And, you can be sure this bidder did not start with quality raw materials because the bidder cannot afford them.

Another Deming principle is that you do not want or need multiple suppliers for one basic raw material. You just need one good one who is—again—your partner. No matter how close two samples measure, two suppliers cannot produce the exact same product. There will be differences that you cannot see that will have an impact on your process. Using multiple suppliers for the same raw material introduces variability that is completely unnecessary. Dr. Deming says no one is "smart enough" to be able to deal with it.

The following recounts a few personal experiences:

EXAMPLE 6.1

I managed a chemical plant for a few years in the early 1980s. One day, I got a call from my boss to come to his office to meet with a new supplier he was considering for one of our basic raw materials. The supplier turned out to be Exxon, and the sales team at the meeting was headed by a vice president. Exxon had just built a new plant to produce the material and wanted to be our supplier. It offered good quality and, of course, an excellent introductory price to help get started in the business. In return, Exxon wanted all of our company's business.

Being who this was, we had little doubt that Exxon could deliver the quality, but my boss was concerned about interruptions in supply because of the new facility. He was responsible for three plants that represented about 15% of the total U.S. market, and he felt we still needed a second supplier as a backup. The Exxon vice president told him he did not need one. He said, "We're Exxon. If something goes wrong at our plant, we'll buy material from our competitor and deliver it to you at the same price." My boss said, "Well, then I guess we don't need another supplier." And we did not. Exxon turned out to be a terrific business partner in every way.

EXAMPLE 6.2

Some years later, I was managing a totally different process at a different company and had two fairly amazing experiences with suppliers. The first one involved an expensive chemical additive. We used the industry standard, which was produced by a well-known, major company (Supplier A). A smaller company (Supplier B) had managed to develop a new substitute that was "almost as good" and cost 10% less. We had run some trials on the new material and concluded that it appeared to be, in fact, almost as good.

Supplier B used to call me from time to time to see if I would be willing to switch in the interest of cost savings, but I had always declined. I had recently been through Deming's training and clearly understood the need for a single, quality supplier. Supplier A had always supplied consistent material at a fair market price, and I really had no idea what the long-term effect of a totally different chemical might be. Then one day, I received a call from one of the plant's senior buyers, who informed me that Supplier A was now selling the standard material to one of our company plants at Supplier B's lower price. She said that the plant was alternating loads from the two companies. I was literally taken aback. I was shocked that anyone would do this. Yes, the sister plant was saving money on raw materials, but at what expense?

I asked the buyer to set up a meeting with Supplier A's senior sales people. A couple of weeks later we met. I opened the meeting by stating that it had come to my attention that Supplier A had been selling its material to a sister plant at Supplier B's price. Supplier A acknowledged that was the case and said it was because it "had a competitive situation" there. I said, "You have the same situation here. Supplier B calls me every other day. However, I tell Supplier B I already have a good supplier that sells me quality material at a fair price." Then I said, "I thought we were partners, and now I find out you're selling the same material to a sister plant at a lower price. I would never do what they're doing. I will only have one supplier, and I'd prefer it was you." The meeting was over. A few weeks later, we received a letter from Supplier A which stated that in the future it would supply our company (five plants) at one price—the same as Supplier B, for a cost savings of $750,000 per year. That was the last we heard of supplier B.

EXAMPLE 6.3

In the same facility as Example 6.2, we encountered a horrendous process failure that resulted in approximately $500,000 in lost sales. We could not produce one of our products and, after depleting our inventory, were unable to ship anything for 2 weeks. The interruption resulted from repetitive, premature failure of an expensive, consumable machine component that was crucial to the process. The component was failing after 1 or 2 days' use when it should have lasted 5 or 6 weeks. As a result, we rapidly went through our spares before we discovered that the components were defective and did not meet a critical specification. And the worst of it was that the manufacturer's documentation certified that, *according to their measurements*, the components met specification. By our measurements, they were off by 50%, so we sent samples back to the manufacturer, who confirmed our findings.

> Again, I asked for a meeting with the supplier's senior sales representatives. At the meeting, I told them that I was greatly disappointed by the inaccuracies in their company's documentation. I also said that, for some time, I had been looking for a backup supplier for this component. We had tried several and concluded that they were all inferior. Since their company had been our sole supplier for years, I considered it to be our partner. One of the things partners do is share in successes *and* failures. Their company's share of our loss was $250,000, but I did not expect it to write a check for that amount. The company could pay it back in future price reductions for its component or however it saw fit. But, in the future, the company would have to provide copies of its manufacturing control charts (per Deming) to confirm that the components were within specifications. The meeting was over. A few days later, we received a letter stating that the company would pay back the loss in free components and comply with the documentation requirement. And, we did not look for a new supplier. We did not need one.

Without Dr. Deming's insight, I would have never understood the partnership a company is supposed to have with its suppliers. The suppliers you need will have a vested interest in your success. But in return, you must be equally loyal as a customer because a partnership only works when it is mutually beneficial. Both partners must be profitable to survive over the years. Occasionally, overzealous purchasing managers can ruin a good relationship by demanding prices that are unreasonably low. In many cases, this is done to advance their personal careers and must be avoided.

Before closing this subject, we must look at one last scenario. It has to do with reducing the quality of a raw material by design to save money. Individuals who further their careers by cutting costs often delude themselves with the notion that they can purchase something that costs less and is just as good. While this may occasionally be possible, it is the exception and not the rule. The overriding concern is the impact the change will have on the variability of the process. If it adds to the variability or makes the process less robust, in all likelihood any material cost savings will be offset by losses in productivity or even quality problems. Yes, we must continue to strive for cost savings, but not if it increases variability and becomes a barrier. The value of material changes must be confirmed with statistical data. If it is in the wrong direction, you must revert to what you know works.

Chapter 7

SMED (Single Minute Exchange of Die)

Introduction

SMED stands for single minute exchange of die. The SMED concept was developed by Japanese industrial engineer Shinego Shingo while working with Toyota in the 1960s. Shingo saw that Toyota's biggest bottleneck was the time it took to change the dies on the large transfer-stamping machines that produced car body panels. The dies were extremely large and heavy, and replacing one took an inordinate amount of time—12 hours or more. So, each die change was a major event and, from a practical standpoint, necessitated long production runs because the time and expense of die changes made short runs uneconomical. This translates into big inventories, more costs, and very large consequences of error.

So, what does this have to do with manufacturers not in the auto industry? The answer is plenty—unless you only make one product in one color on each of your production lines. However, if you make more than one product on a production line, applying the SMED principles should save you a lot of time and money and reduce variability, which is the *main thing*.

The Vision

Were the Japanese ever able to reduce the amount of time to perform an automotive die change from half a day or more to one minute? No. But, they reportedly got it down to 10 minutes. They were able to do that because they had a vision: one minute. You will need to make the same kind of assessment of your operation. The main question is what it is costing you in time, labor, materials, waste, and inventory size to make a product change-over. And, after the changeover, how long does it take after startup to be making first-class quality product again?

Just like the Japanese, you will also need a vision. So, here is one based on Point 9 of our Model Vision. The ideal changeover would be instantaneous: You push a button, and one minute later, instead of making one product, you are now making a different product. And, the new product is immediately good quality. Since this is a common dream among manufacturers, they invented a new term: one touch exchange of die (OTED). The vision for changeover under OTED is 100 seconds or less. If this is not a realistic vision for your facility, you will need to develop one that is—aim high.

Getting Started

In the context of producing multiple products on a single production line, there are some steps associated with a changeover that would appear to be universal and somewhat self-evident:

1. After shutting down for a changeover, the process must be rendered safe for people to enter and work on it.
2. The process should be cleaned.
3. Machine settings have to be changed.
4. Some components of the process may have to be replaced (just like the stamping dies).
5. Some raw materials and operating supplies have to be replaced.
6. After initial startup, additional product sampling is needed to ensure the transition is successful.
7. Machine settings may have to be tweaked before process centerlines are attained.

We analyze each of these in detail.

Step 1: Render the Process Safe

While you would like for your employees to be able to function with the efficiency of a NASCAR pit crew, they have to be able to perform their tasks safely. There is only one condition that provides a sufficient level of safety for people to place all or part of their bodies inside a dangerous piece of mechanical equipment. That condition is that the equipment must be *disconnected* from any power source, and all stored energy must be dissipated. The power sources must then be locked out and tagged with industry-standard danger tags using written plant procedures. A more detailed discussion of locking and tagging can be found in a separate section in Chapter 8.

Step 2: Clean the Process

If you have to shut the process down for a number of minutes to make a changeover, you may as well take advantage of the accessibility and safety the shutdown provides. Now is a good time to blow out or vacuum all those inaccessible areas and wipe everything down. Product color changes will usually make this step mandatory. And, if cleanup is inadequate, the new product may well be contaminated with particles of the previous color or just crud that would have stayed in place if left undisturbed.

Even this cleanup step will require analysis, development of a specific procedure, and a checklist. Over time, you will discover new items to add to the process and figure out what cleanup items are critical. This write-up is another document for the process control manual.

Step 3: Change Machine Settings

The SMED process really begins when you change machine settings. Different products typically require a number of machine adjustments to be made. This may be due to different sizes or different raw materials. To SMED these settings, you will have to be able to reproduce them numerically repeatedly with great accuracy (as opposed to Magic Marker tick marks on a piece of duct tape). By the way, your top operators already know most of these settings and carry them around in a little black book in their shirt pockets. To be more specific, every linear adjustment will need a precision stainless steel scale permanently mounted with a permanent reference mark; every rotational adjustment will need a degree wheel and reference pointer, and so on. If you have to move it, it has to be moved with precision and repeatability. These things are not usually that difficult. Once you explain the requirements to a team consisting of an operator, engineer, and technician, the team should be able to go over the process and determine what is needed.

If the manual measurements are too slow and not accurate enough for your process, there are plenty of high-tech transducers available. And in today's digital world, costs have come down, and accuracy and reliability are higher than ever. The real benefit is that you will be able to see all of these settings in real time on a computer screen and will have a permanent record that is more reliable than visual readings recorded on a paper checklist. The downside to this approach is much higher initial investment and higher maintenance costs. Also, these devices are somewhat vulnerable to damage, and if the damage went undetected it could cause mass product defects.

If at all possible, machine setting changes should be made without the use of external tools. Another term for this is "wrenchless." This means that setting changes are performed with cranks, levers, permanently installed clamps, indexing pins, and the like. When manual positioning devices are too cumbersome, installation of motorized positioners must be considered.

In addition to machine physical settings, there are of course corresponding process variable settings that may change from product to product, like speed, pressure, temperature, current, and voltage. All of these will require another look to be sure each is accurate. Transducers wear out over time, so you cannot take any of them for granted. Each will have to be calibrated and verified to be in good working condition on a recurring schedule with rigorous discipline.

To summarize, *if there is a machine or process setting that must be adjusted, there must be an accurate, numerical measurement to define that adjustment and a rapid, easy-to-use way to reposition the hardware.*

Step 4: Changing Components

Just like the dies that had to be replaced, there are similar components in other processes. In the paper industry, there may be fabric and roll changes for different grades of paper. Mold changes are routine in the injection molding industry, as are polymer changes for different parts. It should be obvious that all stamping and forming processes have parts that must be replaced in product changeovers.

Sometimes, replacement of the individual components that define the product is so complex and time consuming that it is much easier to roll in an entire machine section and bolt it up in a matter of minutes. These are referred to as “quick change assemblies.” Top fuel drag racers use these more than anyone else. Apparently, they can rebuild an entire supercharged engine between runs in 6 minutes.

Another important advantage of quick change assemblies is that they can be meticulously assembled in a shop without the pressures and time constraints of a product changeover. And for that reason, quick change assemblies should be used as much as possible because they should be more reliable and have a positive effect on variability.

The obvious drawback to quick change assemblies is that they can be quite expensive. You may have to build an entire rolling machine section to make it worthwhile. The trade-off is lower downtime and potentially reduced startup waste versus increased cost for the added parts. How often you are required to make changeovers and the amount of time it takes will have to be weighed against the cost of the quick change assembly.

Whether or not you use quick change assemblies, *all* of the parts that must be replaced in a changeover must be kept segregated in kits and meticulously inventoried and maintained just like the quick change assemblies.

Step 5: Changing Out Raw Materials and Supplies

Different products obviously may require different raw materials and operating supplies. How rapidly and efficiently your team is able to replace them is the part that is SMED related. This item is more like the activities of a NASCAR pit crew than the others; the following are some of the things to study:

- Where and how materials are staged
- Material loading fixtures
- Lifting tools
- Numbers and placement of people
- Motion–time studies of individual operations

When we get into the SMED development process, we discuss some of this in more detail.

Step 6: Additional Sampling

In a perfect world, when a process is restarted after a changeover, it would make good quality product from the first piece produced. Since this is unlikely, product quality will have to be closely monitored using increased sampling frequencies to ensure that nothing went wrong.

Step 7: Final Adjustments and Centerlining

Now, the process is back up and producing acceptable product, but during the transition period, a number of machine settings were tweaked. Those final settings become the new process centerlines. In other words, day in and day out, these settings should correspond to products that meet specifications. It is critical to capture those final settings in a log and compare them to previous centerlines. If your team has been doing this correctly, over time, the number of final adjustments will be reduced along with a corresponding reduction in waste. Eventually, the need for tweaking will all but disappear. Steps 6 and 7 are where the Shewhart PDCA (Plan, Do, Check, Act) cycle comes into play repeatedly.

The SMED Process

While a lot of SMED is self-evident, books have been written on the subject, and you will want your in-house expert to be trained on all of the various aspects that are beyond the scope of this book. An example reference is Shigeo Shingo's 1985 book, *A Revolution in Manufacturing: The SMED System* (Productivity Press: New York). However, for our purposes, the points we deal with are the basic concepts managers need to be versed in to direct the SMED effort in their facilities. With that said, here is one way to implement SMED:

1. Assemble a SMED team and determine the need for SMED. This is a purely mathematical exercise. You will need to know the number of product changeovers you have, the time it takes you to make a changeover, and the amount of waste and delay associated with each one. That dollar amount is the opportunity available and is the basis for the true SMED vision for your facility. You will not be able to get it all, but you should be able to save at least half of that, including the investment in time and equipment needed to modify your equipment. If nothing else, implementing SMED will inherently reduce variability.
2. Your SMED team must be cross-functional and include operations, maintenance, engineering, and finance. And to be effective, the team will

need plenty of training. One member of the team must be appointed overall leader. That person should have broad experience, solid technical skills, and be a long-term employee as opposed to a youngster who is looking to be upwardly mobile or the like. My personal pick would be a seasoned engineer who is more or less "permanent." It is recommended to make the SMED assignment a promotion for the leader to emphasize that you are serious and have high expectations.

Last, commission your team officially by providing a "charter," which contains your expectations and the team's level of authority, including the amount of money the team is authorized to spend. This spending authority demonstrates more clearly than anything else that you believe in the team and that the team is empowered to make decisions for you. The commissioning "ceremony" should be done at a weekly leadership meeting and an announcement posted on official bulletin boards. This communicates the team's authority to act and its endorsement by the management team.

3. The initial assignment for the SMED team is to calculate the financial opportunities and prioritize them on a Pareto chart. You cannot implement SMED on everything at once, so you will need to select one production line to start as a pilot. During this pilot phase, the management team must be involved and supportive.
4. The next step is to assess the specific production line. The SMED team must attend crew meetings and give an overview to all those working in the area. This step is crucial. Unless everyone is involved and trained, the project will fail. Besides, you want everyone's ideas and suggestions. Answering their questions will force the SMED team to think through some things not considered up to that point.
5. Decide when the next major product changeover is scheduled and inform the crews that the SMED team will be videotaping the entire process. Buy several digital video cameras—they are cheap—and hand them out to the team. Video every aspect of the changeover. Talk to people while you are doing it and ask them what they are doing and why. This includes mechanics, operators, everyone. Capture the positions of people, tools used, and procedures.
6. The SMED team must, as accurately as possible, record the time required to perform the changeover before implementing SMED. There are two components: the time the operation is shut down until the time it is restarted and the time from restart until good-quality product is made at normal production rates.

7. The SMED team next reviews all the videos and edits them down to the essentials so they are suitable to share with crew members. At this point, the videos should be chronological and tell a story.
8. The SMED team then schedules an extended "brainstorming" meeting with selected crew members. This will take quite a bit of time, and if several sessions are needed, so be it. Do not rush this part. At these meetings, go over each video clip and ask the following:
   - *What* is this person doing?
   - *Why* are they doing it?
   - *How* could it be done *better*?

   In initial SMED meetings, it is common to find whole process steps that can be eliminated by merely altering a sequence. Also, there should be many opportunities to employ new, powered lifting and loading equipment and improve the ergonomics of every job. Communicate the objective to make the changeover without external tools (wrenchless) and solicit ideas for what is needed to make that possible. Encourage "blue sky" thinking—ideas without regard for cost or practicality.
9. The SMED team processes the feedback they obtained from the crew and puts together an action plan. The team should use the NASCAR pit crew mindset. If extra people would speed up things, try bringing in some on an overtime basis and see if there is a net financial benefit. *Consider using maintenance technicians and even managers and other office people to help.* In every case, "think out of the box."

   The action plan should have three major headings:
   - Lockout and tagging: who, when, and the procedure used
   - A complete listing of all the process settings that are to be changed
   - How each setting is performed today and what numerical reference and adjustment device (crank, lever, etc.) will be needed to accurately measure, reposition and lock down the new setting
   - A listing of each operation step with the following:
     - How many people required, where they are positioned, and what (if any) additional personal protective equipment they need
     - Lifting and loading equipment required
     - Materials and supplies removed and where stored
     - Cleanup points
     - Location of prestaged changeover materials and supplies
10. With the new action plan implemented, you must now *train the entire crew* on the new procedure. Old ways die hard, questions will be asked, and some of them will shoot a few holes in your plan, which

you will have to fix before proceeding. In any case, this is all good. In fact, it is the best.

11. What do you do next? You *try* it. Get out the video cameras and again videotape everything to see if there are improvements. And—most important of all—record the shutdown to restart time and the time required to make good product at normal production rates. If these are improved, congratulations. Celebrate with the crews, show your appreciation, and again get their inputs for improvements. Use that Shewhart cycle repeatedly and wring out all the efficiency that is practical to attain.
12. After some successes are logged, the SMED team must accurately document the finalized procedure in detail. This must include photos of people and positions, close-ups of settings—everything needed for a new team member to be able to reproduce. This document is stored electronically in a new SMED folder, and a hard copy is put in the process control manual for day-to-day reference.
13. Once the SMED team is satisfied with the results of the pilot, the team presents its findings in a formal presentation to the management team and shows clearly what was accomplished. The team should include an edited video to show management before and after clips. There should be measurable savings in downtime, labor, and waste, and, yes, variability. Make it fun. Make it exciting.
14. If positive results are achieved, it is *incumbent* on the management team to be demonstrably appreciative. After all, this team, with a very small investment, has in all likelihood come up with significant cost savings and quality improvement.
15. At this point, the SMED team should also be prepared to share with the management team the rollout plan and projected timeline. This timeline and progress report become an item for the facility monthly report. If the SMED team is not allocated the time and resources to proceed, all will know. However, done properly, SMED will be institutionalized and become one of those "this is how we do things here" items.
16. Over time, changeover times will be reduced if accurate records are kept and the PDCA process is rigorously followed. The single SMED test data file will evolve into a spreadsheet that contains empirically derived settings for all products on all production lines. Because they are so critical, change access to this file must be restricted to the SMED team.
17. On rare occasions, when the planets are perfectly aligned, it will be necessary to deviate from one or two of the standard settings just to make the process run correctly. Due to luck, those adjustments might

just happen to cause the line to perform better than expected. It is imperative that these deviations are captured separately as "good run settings"; that is, they caused the line to run better than normal. It is unknown why they worked that day. It may have been due to an unseen raw material variation or a change in atmospheric conditions. In any case, they are not ready for unrestricted use until they have been verified, and to do that they must be repeatable on more than one run. Then, once their validity is established, they become the new starting point if, and only if, they are approved by the SMED team.

Back to the Vision

The process described is an example of what could be called a "manual" SMED application. It is pretty low tech, and it is also very inexpensive to implement. Do not misunderstand; it is still quite good and has a very high potential to save money and reduce variability. However, our vision is much more ambitious and is actually the one-touch version or OTED. On the downside, OTED happens to be quite expensive. It entails having servo-mechanisms with digital transducers for position feedback on *every* process adjustment point. In other words, every adjustment point is powered and every position is seen in real time on a computer screen. Changeovers are done by typing in a product number and pressing the "Enter" key on the process control computer.

So, is this practical? Is it worth the money? The answer is "it depends." Lean manufacturing implies low inventories and rapid changeovers to fill orders as they are received. This is in contrast to building large inventories of your entire product line and storing them in warehouses. The advantage of the latter is rapid customer service. The disadvantage is all those inventory costs and consequence of error. So, if you could fill orders at will, "on the fly," having this capability might be worth the investment. The sweet spot for your manufacturing situation might well be a blend of the two—some manual settings and some automated.

Summary

Again, while entire books have been written on SMED, the intent of this chapter is only to give you an introduction along with a practical, hands-on method to try some of the concepts. And, one thing to keep in mind if you decide to try it is that you will need a SMED champion to keep the fire burning in your facility. These days, people have more than enough to do in their jobs. Unless you assign the resources and allocate the money, SMED will not be successful. It will be relegated to the dust heap along with the other programs of the month. But done properly, and driven down into the organization, SMED can reduce variability and save you a lot of money.

Chapter 8

The Process Control Manual

Introduction

The process control manual (PCM) is an *authoritative* compilation of all the information an operator needs to know to be able run a process safely and within established specifications. It covers the entire operation from start to finish and is the basis of all operator training. It is the foundation for Point 4 of our Model Vision—training our workforce to be highly skilled professionals—which is our answer to Deming's Point 6. But, more important, it is another key to reducing variability because *its primary purpose is to ensure that everyone does everything the same way.*

To be authoritative, this manual must be kept as current as possible. In that context, all process or product changes must be incorporated into a manual revision prior to implementing the changes in the plant. Also, formal, periodic reviews are mandatory. To make that visible to users, the first section of the PCM contains a running log of reviews or revisions, the nature of the change, and the signature and date of the person who did the work. To institutionalize the review process, the plant's electronic "tickler" system should have a complete listing of all manual sections and when review dates are due (it is impractical to review an entire manual at once) and should automatically send notices to production teams. The tickler should be answered in writing by the department manager, certifying to the plant manager that the manual has been reviewed and is now up to date. A summary report of overdue manual reviews should be sent to the plant manager.

The original documents for PCMs must reside in the plant's computer system. This makes all the plant's manuals accessible to all employees.

Department managers are free to see how different departments perform operations and can adopt what they feel is the best methods. Manual sections, checklists, and diagrams can be printed out by employees and studied for their personal training and daily use. The one thing to watch for here is security. Access to print proprietary information will have to be set up with some levels of authorization.

The PCM should contain the following sections as a minimum:

1. Products or product components produced
2. Product specifications
3. Process description
4. Process diagrams
5. Raw materials used
6. Specific operating supplies
7. Detailed operating procedures for each job position in the process
8. The SMED (single minute exchange of die) procedures for this particular operation with a sample settings data sheet
9. Process readings sheet for parameters that are not computer monitored
10. Required tool list
11. General safety procedures
12. Locking and tagging procedures
13. Housekeeping checklists
14. Statistical process control methods used
15. Product attributes

As usual, we discuss each of these in detail.

Products Produced

This section is an introductory narrative that starts with a description of the product and how it is used by the customer or in the next step in the manufacturing process. The intent is to show employees why their specific job is important and what the consequences of error are. Explain the impact specific defects cause to establish clearly why these details are critical. This is also a good place to introduce the "customer" concept. In other words, the next person who receives the output from their operation is *their* customer, whether they are inside or outside the plant. For some reason, customers inside the plant do not appear to merit the same consideration as those outside the plant. But in reality, they are equally important, and this concept must be taught to each employee.

This section ends with a sample table of product codes the line produces but with a note referring the reader to obtain current data from the process control computer. While this table should be kept up to date as much as practical, it is there only for illustrative purposes and must be clearly flagged as such.

Product Specifications

A complete set of the latest official product specifications should be filed under one tab of the PCM. While the originals are maintained in the plant's computer system, the copies serve as a handy reference and training aid.

Process Description

The narrative next moves to a detailed description of each processing step from introduction of raw materials to the finished item. It should "tell a story" of the process flow that includes manufacturers and model numbers for each major piece of equipment as they are encountered in the process along with some pictures from the manufacturers' tech manuals. Photographs of selected close-up details can also be placed to add more depth and interest. This section should be such that a walking tour will bring the manual pages to life.

Particular attention should be paid to each safety feature incorporated into the process, with each one described in detail. In addition, provide a diagram and photos to show the locations of emergency stop buttons, eyewash stations, safety showers, and other items that are specific to a machine or operation step. An example diagram is shown in Figure 8.1.

Process Diagrams

Process diagrams should start as an overview in simple block flow format and progress to greater depth and detail on subsequent pages. Drill down into each of the blocks and create separate flow diagrams to show what goes on inside. For continuous processes, like chemicals or pulp and paper, it is

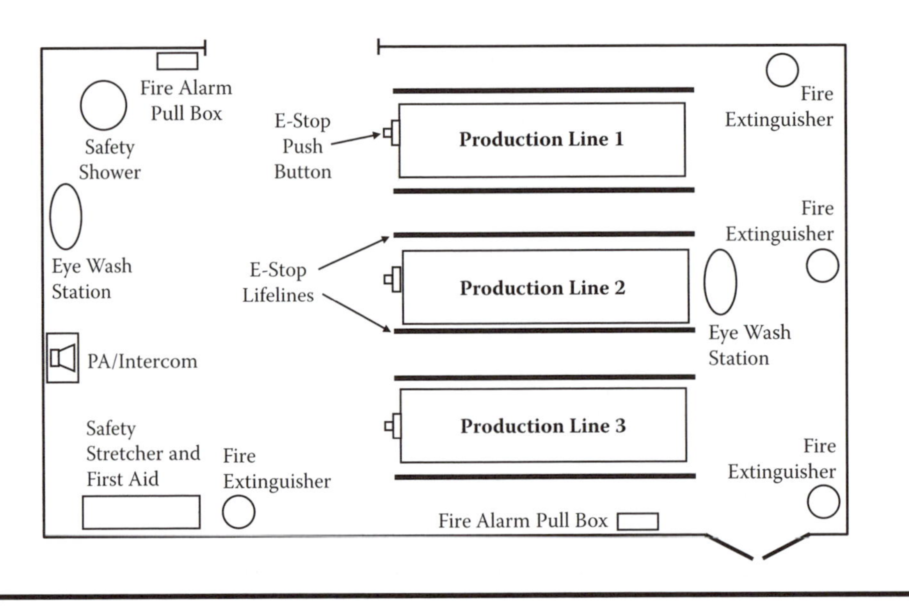

Figure 8.1 Safety equipment locations.

mandatory to include process and instrument diagrams (P&IDs) that show control elements like valves, pumps, and tanks to convey how the control logic works. The function of all safety devices, interlocks, and control overrides must also be included. These diagrams are an extremely important part of the manual and should over time be committed to memory by operating and maintenance team members. An example P&ID is shown in Figure 8.2.

Raw Materials Used

All currently approved raw materials should be listed in a table by manufacturer and manufacturer's product number so that operators have a ready reference to verify that the materials on the floor are correct. Each specification must include the manufacturer's safety data sheet.

A spreadsheet of raw materials and product code numbers will assist operators in comprehending the "big picture" of how products are made. As before, source documents for this critical information must reside in the plant's central computer system. Where products require a "recipe" list of different materials, a sample recipe form should be printed and placed in the manual for training purposes.

It is critical that a computer log be established to record all raw materials used in a production run. As a minimum, the log should record manufacturer, product number, lot number, and time introduced into the product. If

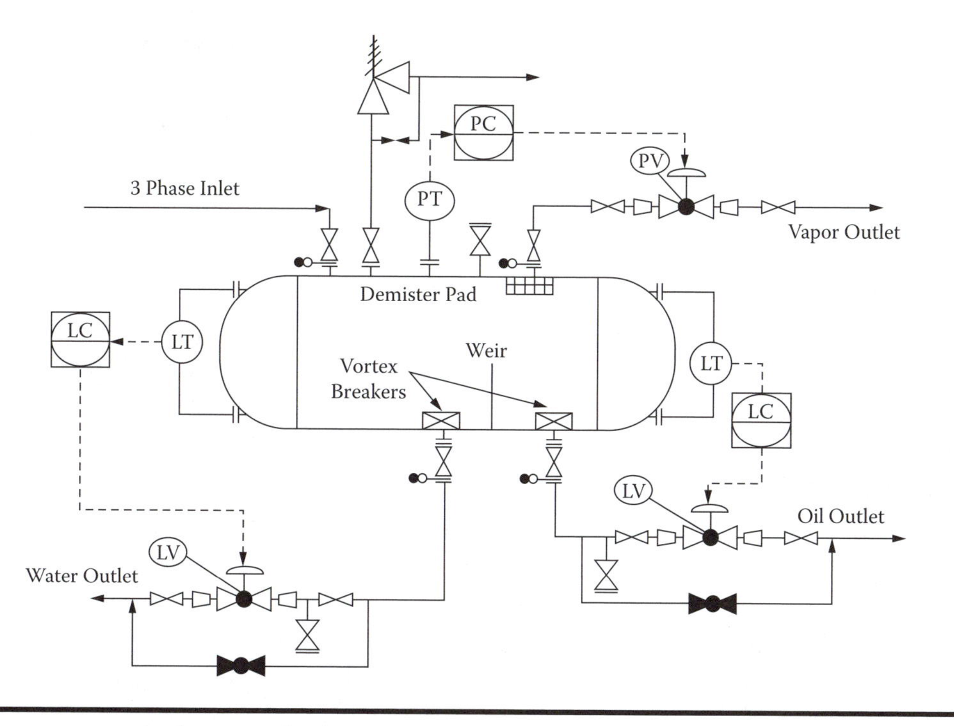

Figure 8.2 Example P&ID drawing. (Courtesy Engineering Design Encyclopedia, www.enggcyclopedia.com.)

the materials are used in bulk and piped in, then the plant needs systems to correlate material delivery dates to production runs. Variations in bulk materials can be disastrous, and this is why it is so important to be able to trust suppliers. Deming tells us that a quality supplier will be able to provide copies of statistical process control (SPC) data for each lot produced. If you want them, you will have to ask for them. Do not ask for them unless someone is assigned to review them.

Authorized Operating Supplies

Operating supplies are all the incidental items needed to manufacture a product. They are not raw materials or packaging supplies. They are things like cleaning solvents, wipes, tape, lubricants, special tools, and personal protective equipment.

Obviously, anything that comes in contact with your product must be scrutinized carefully. Solvents can damage the finishes or leave residues and accompanying odors; the same is true for lubricants. As stated previously, junk in a plant increases spontaneously, or at least it seems that way, and operating supplies are no exception. Occasionally, even dangerous, highly flammable solvents may show up unexpectedly (because they always seem to work the best). The only way to ensure the *prescribed* operating supplies are there is by management audit.

Personal protective equipment also multiplies because, after all, it is “free,” so people tend to accumulate plenty of spares. Also, your team must decide exactly which protective equipment is best suited to the specific tasks done in an operation, and that is what should be specified. Other types may be inadequate or cumbersome. A good example is respirators. There are many different types available, but the best one is the one that is *tested* to meet a specific environmental condition that is also the most comfortable and least intrusive.

To summarize, your PCM must have a section that lists *exactly* what operating supplies you and your team approve, including allowable quantities. Deviations from the list should be noted and immediate corrective action taken when management or quality technicians make routine walk-throughs. Keep this activity as “light” as possible. Provide gentle, patient coaching and do your best to explain why a material is not permitted. Do not be arbitrary. People are all the same, and just like your kids, they would like to know why something is not permitted.

Detailed Operating Procedures

There is a safest, best method for performing any operation. For that reason, operations like startups, shutdowns, or loading raw materials must have detailed, written operating procedures that have been tested and retested through years of continuous refinement. Correspondingly, employees need to be trained and retrained to use them until they become second nature.

Procedures with the biggest consequence of error or that contain a large number of steps must be in checklist form to ensure each step is carried out in the prescribed sequence. This will prevent unnecessary production delays and possible equipment damage. Airline pilots face huge consequences of error, so they use checklists. Similarly, there are manufacturing processes that will explode or go through catastrophic failure if operated incorrectly. If for no reason other than SMED, there is a clear need to document, audit, and update operating procedures.

Every time a procedure is used, employees should be strongly encouraged to point out any detail that they find incorrect. The simplest pencil notes on the procedure form are all that is needed. Examples are "Wouldn't start," "Stuck—had to use wrench", "Need to do _ first." Then, as soon as the dust settles and there is more time available, these notes must be discussed with the employee in detail to find out what actually happened. First and foremost, get the employee's ideas on how to remedy the problem and then call in whatever other resources are needed to develop a permanent solution. Regardless of what it takes, each bump in the road must be worked on by the team until the procedures work smoothly. This is another application for the Shewhart cycle.

SMED Procedures

In reality, SMED is another operating procedure, and for that reason a copy is kept under a separate tab in our PCM. Since the originals are maintained in the process control computer, employees can print out copies as needed. Prior to a changeover, the team typically will hold a brief meeting to go over the playbook one last time before the kickoff. Handing out paper copies makes the meeting more effective and provides a convenient place for team members to make note of any hiccups they encounter.

Process Reading Sheet

If all of the process variables in your operation were monitored by computers, a human would not need to look at them very often because the computer can be set to alarm 24-7 whenever any anomaly is calculated. Few of us have that luxury. While some variables may be computer monitored, a number of them will not be, and those will need a visual reading sheet. You will have to decide for yourself what frequency is appropriate. Needless to say, these readings should be kept to a minimum to avoid unnecessary labor and drudgery. An example reading sheet is shown in Table 8.1.

The real purpose of the readings is for a human to look at a gauge and make a judgment *at that time* regarding whether the reading is within normal limits or determine if an unfavorable trend is developing. Therefore, the reading sheet must contain sufficient background information or the activity is meaningless. Another visual aid is to print the allowable upper and lower limits on adhesive-backed labels and place them on all the gauge faces. Then, *anyone* walking through an area can spot a gauge out of range.

Table 8.1 Process Reading Sheet

Date				*Allowable Range*		
Item Number	*Description*	*Time*	*Current Value*	*Low*	*High*	*OK: Yes/No*
1	Chiller inlet temp (deg F)			85	100	
2	Chiller exit temp (deg F)			40	45	
3	1st press pressure (PSIG)			100	115	
4	2nd press pressure (PSIG)			200	225	
5	3rd press pressure (PSIG)			300	330	
6	4th press pressure (PSIG)			400	450	
7	Framostat position (% open)			60	80	
8	Kanuder valve (open-closed)			Closed	—	
9	Recirc pump inlet press (PSIG)			10	20	
10	Recirc pump dischg press (PSIG)			60	80	
11	Air supply pressure (PSIG)			100	125	
12	Steam supply pressure (PSIG)			300	350	
13	Plenum differential (in. H_2O)			6	8	
14	Main drive speed (RPM)			1,000	1,800	
15	Main drive current (amp)			200	600	
16	Nurdlinger pump speed (RPM)			1,500	3,250	
17	Polymer tank level (inches)			30	70	
18	Agitator torque (in.-lb)			240	610	
19	Wash water pH			6	8	
20	Wastewater conductivity (mMhos)			80	100	
21	Flash tank overflow (yes/no)			No	—	

Because no one really wants to pore over a stack of reading sheets every day, they are often ignored. Eventually, operators may conclude that no one cares about them and will begin to fill them out in the break room. The moral here is that if you are going to pay someone to take readings, then appoint someone to take the time to scan them and note any deviations from standard.

Required Tools

As stated in Item 6, there are preferred methods for performing operation steps. Those steps may involve the use of specific tools; therefore, those tools must be available at all times. The preferred way to store tools is on a Peg-Board with a painted outline and label for each tool: "a place for everything and everything in its place." Small, incidental tools can be kept in toolboxes provided that they are inspected often to verify that what is in there is authorized in the PCM and in good working order.

General Safety Procedures

Here are some things you should find in the general safety section of your PCM:

- General safety rules like removing ties, rings, and jewelry; getting rid of chewing gum, and the like
- Personal protective equipment required to enter the department
- Map of machine emergency stop buttons
- Map of emergency showers and eyewash stations
- Map of emergency exits
- Map of fire extinguisher types and locations
- Map of fire alarm boxes
- Intercom/public address (PA) locations

It may be possible to display some of these safety measures on a composite map. However, if it becomes too cluttered, the map loses its visual effectiveness as a training aid. One way to deal with that is to print each category of safety items in different colors on transparencies so that they can be superimposed.

Locking and Tagging Procedures

Perhaps the most important documents of all in the PCM are the locking and tagging procedures. Each machine in a plant has differences regarding how it is deenergized and rendered safe, so locking and tagging procedures must be itemized for each machine. There are also different levels of lockout: machine, line, department, and plant. Procedures to cover each level are also mandatory.

Failure to completely deenergize production equipment can result in loss of life and limb. As a manager, this is your highest responsibility. *You* must ensure that all energy sources are *disconnected* and all stored energy in pneumatic and hydraulic devices is *dissipated*. Hot surfaces must be cooled, and stored electrical charges must be grounded. Since you personally cannot be expected to oversee the work, you must have bulletproof systems in place that do this for you—throughout your organization—24-7.

After removing all energy sources, locks and tags must be placed to ensure that the equipment cannot be inadvertently reenergized. The final step is that start buttons and switches must be tried to verify that the system is safe. Then and only then can people be allowed to work on the equipment. (This is referred to as the lock, tag, and try process.) There is only one way to ensure that this process is carried out correctly: with a written checklist that is signed off by the *trained* individuals performing this task.

The preceding discussion is the condensed version. Lockout is a bigger subject that needs to be studied by your in-house safety expert, engineering, and maintenance teams. Lockout procedures should be reviewed with *all personnel* at least annually in safety meetings.

Housekeeping Checklist

Just like your home and garage, there are recurring cleanup chores that require daily attention. Crud accumulates in certain places on the machinery, trash barrels have to emptied, floors need to be swept, and so on. If you expect to keep all of it clean every day, you will need a list to assign specific responsibilities to each crew position. Each day, the crew leader or supervisor should go over the list and make sure his or her crew leaves the department in good condition. As we discussed in Chapter 2, the only way you can keep this going is if management makes walk-through inspections. When doing these, if you see something that is out of place, always ask a question like, "Do we have a storage place for that?" or "Does this floor meet our housekeeping standards?" Do not nag. People are busy, and nagging gets old fast. Just let them know that you know that the place is not up to standards, go away, and give them time to clean it up. Then, go back and say, "The place looks great! Thanks."

Statistical Process Control

Finally, we have covered all of the prerequisites in our PCM to enable us to talk about its main purpose: SPC. We discussed at some length what is required to implement SPC in Chapter 5 and do not repeat that here. In this section, we focus strictly on sampling, product evaluation, and control charting of product variables. Product attributes are discussed in their separate section.

Product Sampling

This section has one purpose: to define when, how, and how many items are to be pulled from the production line for sampling. Since literally everything depends on performing this step without bias, everyone in the department must be thoroughly trained to do it properly. As stated previously, the best way to do this is by machine to remove the human element completely since the machine can be programmed to kick out samples using a random number generator algorithm. In contrast, a human does it by the clock and must work at being unbiased.

If your production line makes a steady stream of products but moves relatively slowly, a person should be able to remove samples manually. Designate a safe location at the end of the line just before the product is placed in the shipping case. If the cases are packed by machine, you will need to remove samples from finished cases to ensure that they are not damaged by the case-packing machinery.

If your process is too fast for a person to remove samples, an ejection system will have to be installed to kick out the samples. While you are installing this hardware, you might as well make it automatic and random. Make absolutely certain the ejector and collection hardware do not damage the samples. If all of this is too problematic, you will have to resort to case sampling. (Case sampling is a pain. You have to cut open the cases, collect random samples, and manually put the case back together with new product.)

Remember that it takes a minimum of three to five samples at a time for the results to be statistically significant.

If your process makes a relatively small number of pieces per shift, say 100 to 1,000, the entire shift's production must be isolated *in one location in the order produced*. You will need to pull a minimum of five randomly selected samples for inspection. Use a random number generator to select the points in time the items were produced.

Product Evaluation

OK. You have the samples. Now what?

In Chapter 5, we performed a survey to define what product variables *and attributes* we needed to verify and make a report. This section of the manual is devoted to specifying in unrelenting detail how every measurement is performed. You will need high-quality, professional-grade calibrated lab instruments to ensure accuracy. Remember: these measurements become the feedback information used to make adjustments to process settings—*they have to be repeatable.* In addition, they can be measured by the customer and will be if they are critical to the customer's business.

Once the measured values are collected, the lab technician enters the X-bar and range numbers into the process control computer in the lab, and viola, the updated control charts are displayed in the plant.

Control Charting

The samples have been pulled, the product tested, and the control charts updated as shown in Figure 8.3. Finally, the moment of truth has arrived. The question is, what will your operators do with the information? The answer is—*It depends.*

Definition of "In Statistical Control"

Remember, you cannot even begin to use control charts unless you have removed all known sources of process upsets and your plant is running smoothly. We did that in Chapters 3 and 4 with maintenance and automatic controls. Also, we verified our work by performing capability studies in

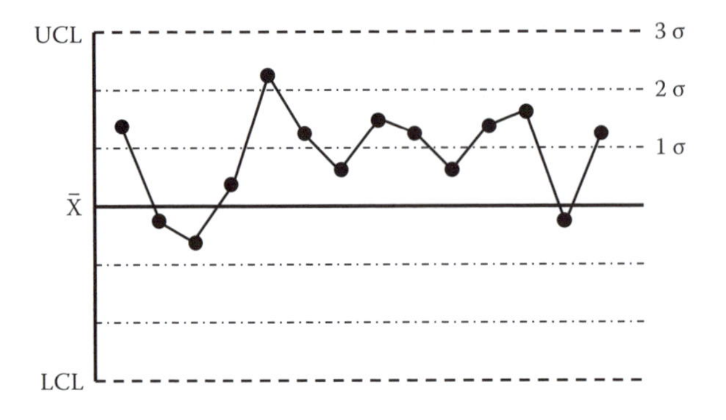

Figure 8.3 Example control chart. UCL = upper control limit; LCL = lower control limit.

Chapter 5 and confirmed that all the "red beads" we could find were eliminated from our system. In other words, the process is now *in statistical control*, which defined as the

> state of a stabilized production process in which only common causes of variation remain (all special causes of variation having been removed), as evidenced on a control chart by the absence of (1) data points beyond the control limits, and (2) non-random patterns of variation. (BusinessDirectory.com)

The keywords are italicized here: "*the absence of* (1) data points beyond the control limits, and (2) non-random patterns of variation." What this means is, left to itself, the process should be able to amble along within plus or minus three sigmas with little or no assistance from an operator.

Recognizing when your process has achieved statistical control is vital. So, rules for identifying out-of-control conditions were first postulated by Shewhart, who invented the control chart around 1920. Decades later, more formal rules were published in the October 1984 edition of *Journal of Quality Technology* by Lloyd S. Nelson. Nelson came up with the eight rules shown in Figure 8.4. As you can see, some of them are subtle and take an expert to interpret.

Real Purpose of Control Charts

While control charts provide useful information about the state of our operation, their main purpose is to tell us *when it is appropriate to make an adjustment* and avoid the natural tendency to overadjust, which greatly increases variability. There are established rules for when an adjustment is appropriate. These are provided for Statit® software in Appendix 2.

Rather than force operators to agonize over whether they should make a process setting adjustment, we have the luxury of letting the software make that decision for us. This removes the variability of individual operator interpretation and is also a real time saver.

Back to the Process Control Manual

All of the information needed to train operators to do control charting and adjustments must be included in this section of the PCM. It must contain example control charts that are in control and others that are not and why

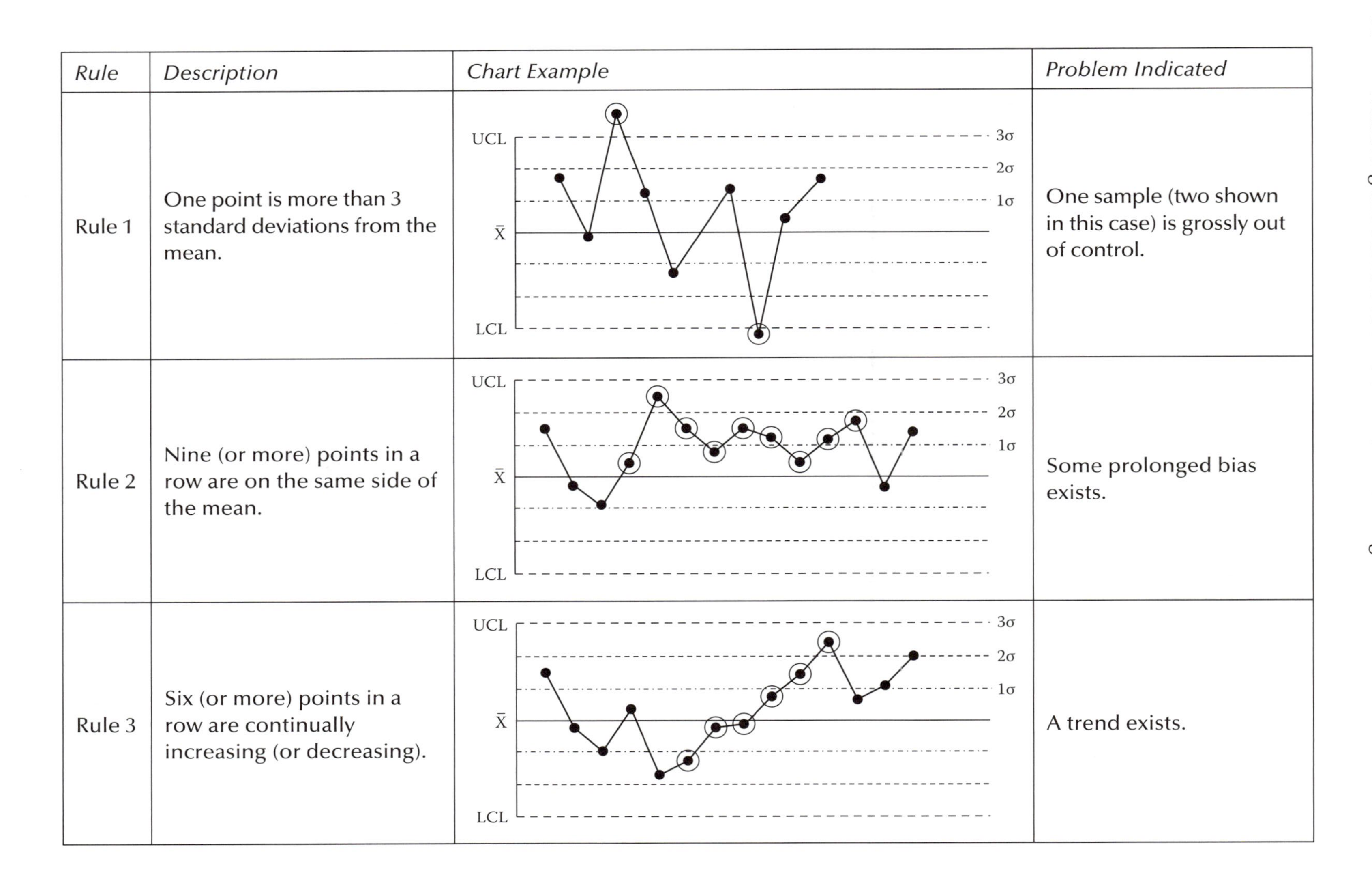

Rule	*Description*	*Chart Example*	*Problem Indicated*
Rule 1	One point is more than 3 standard deviations from the mean.	UCL, $\bar{X}$, LCL; 3σ, 2σ, 1σ	One sample (two shown in this case) is grossly out of control.
Rule 2	Nine (or more) points in a row are on the same side of the mean.	UCL, $\bar{X}$, LCL; 3σ, 2σ, 1σ	Some prolonged bias exists.
Rule 3	Six (or more) points in a row are continually increasing (or decreasing).	UCL, $\bar{X}$, LCL; 3σ, 2σ, 1σ	A trend exists.

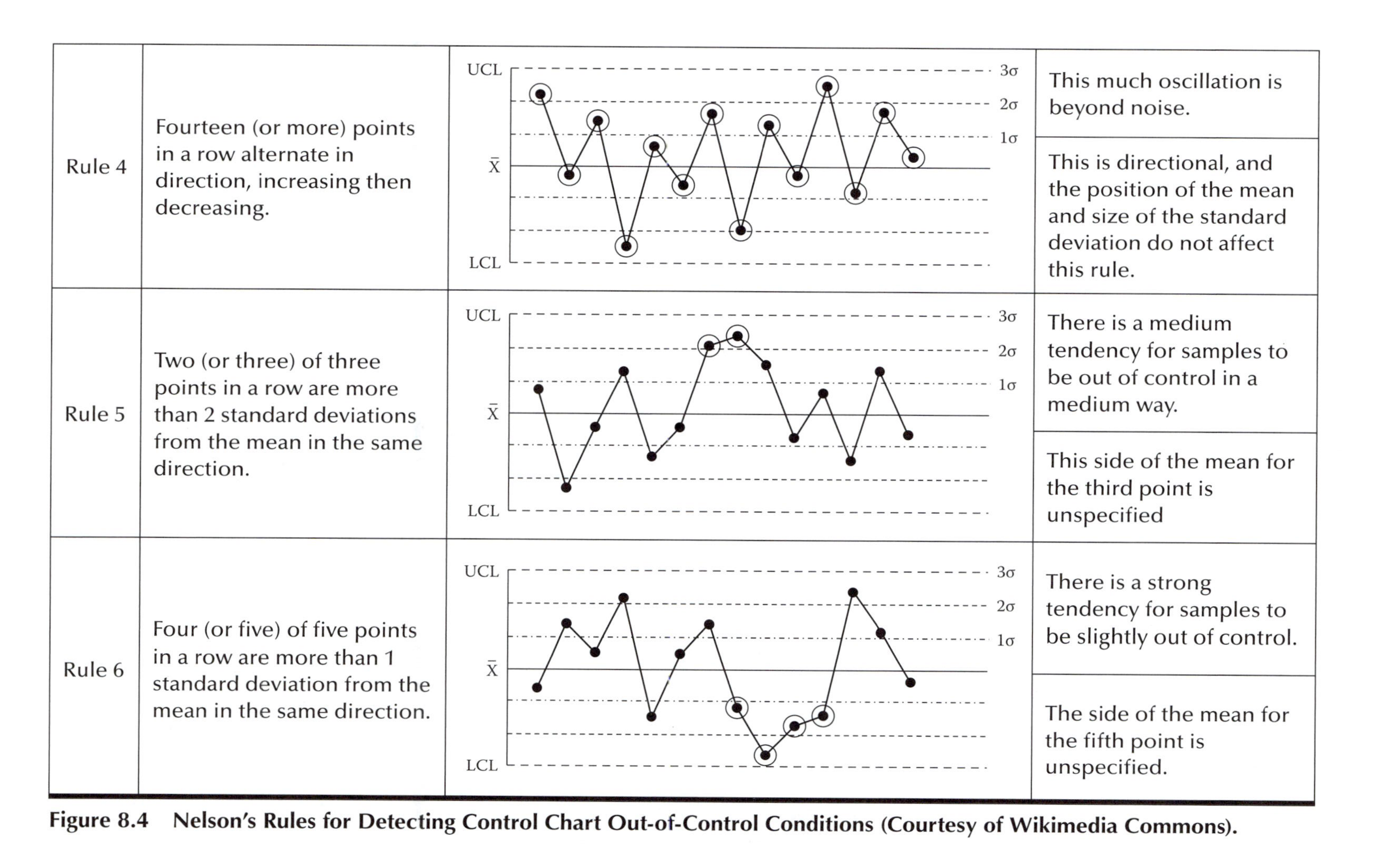

Rule	Description	Chart	Notes
Rule 4	Fourteen (or more) points in a row alternate in direction, increasing then decreasing.	UCL, $\bar{X}$, LCL, 3σ, 2σ, 1σ	This much oscillation is beyond noise. This is directional, and the position of the mean and size of the standard deviation do not affect this rule.
Rule 5	Two (or three) of three points in a row are more than 2 standard deviations from the mean in the same direction.	UCL, $\bar{X}$, LCL, 3σ, 2σ, 1σ	There is a medium tendency for samples to be out of control in a medium way. This side of the mean for the third point is unspecified
Rule 6	Four (or five) of five points in a row are more than 1 standard deviation from the mean in the same direction.	UCL, $\bar{X}$, LCL, 3σ, 2σ, 1σ	There is a strong tendency for samples to be slightly out of control. The side of the mean for the fifth point is unspecified.

Figure 8.4 Nelson's Rules for Detecting Control Chart Out-of-Control Conditions (Courtesy of Wikimedia Commons).

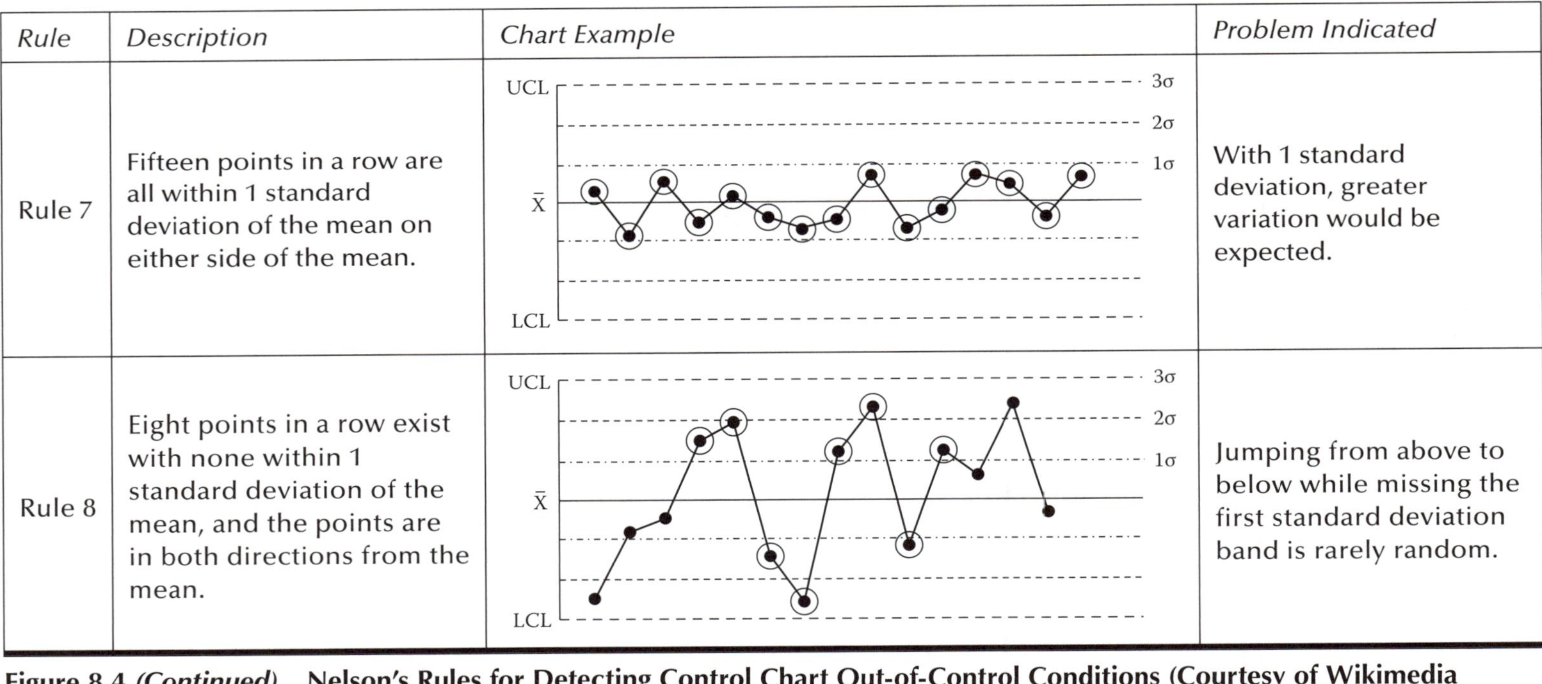

Rule	*Description*	*Chart Example*	*Problem Indicated*
Rule 7	Fifteen points in a row are all within 1 standard deviation of the mean on either side of the mean.	UCL, $\bar{X}$, LCL; 3σ, 2σ, 1σ	With 1 standard deviation, greater variation would be expected.
Rule 8	Eight points in a row exist with none within 1 standard deviation of the mean, and the points are in both directions from the mean.	UCL, $\bar{X}$, LCL; 3σ, 2σ, 1σ	Jumping from above to below while missing the first standard deviation band is rarely random.

Figure 8.4 *(Continued)* Nelson's Rules for Detecting Control Chart Out-of-Control Conditions (Courtesy of Wikimedia Commons).

based on the rules. The operators should have sufficient information *and training* so that, even if the computer system crashed, they could keep running manually with minimal risk. Having this level of insight is invaluable and avoids making knee-jerk adjustments without taking into account other plant conditions. It also reduces the likelihood of people blindly operating "from the neck down."

Product Attributes

While variables can be measured, attributes are usually subjective. In other words, they must be manually and visually compared to validated standards. Examples of attributes are finish gloss, smoothness, softness, cleanliness, appearance, and print quality. For each of these, examples of good (acceptable) and bad (rejectable) product must be available for comparison; otherwise, it will be impossible for a quality technician to make rational decisions. If possible, place photos of good and bad attributes in the PCM so that people on the operating floor understand what the goals are and be able to head off problems in the early stages.

While sometimes difficult to do, it is highly desirable to have available a range of attribute standards. In other words, you will have an example of the best product you have ever made at one end of the spectrum, one that is defective at the other end, and several graded versions in between. It may take some concerted effort to collect these specimens, but over time you will be able to find them. *Once you have them, you can assign numerical values to the attribute's measurement.* And, the numerical values can be entered into the process control computer on run charts.

Note that it is possible to control chart attributes and even percentage defective. Statit software has that capability. However, those are beyond the scope of this discussion.

Summary

That concludes this section on the PCM. It is not just a book lying around gathering dust. It is the source of all process knowledge and the training textbook for your operators. They must be able to rely on it without question. Inaccuracies in the PCM make it a liability rather than an asset and reflect unfavorably on your entire organization.

Chapter 9

Training and Retraining

Introduction

Insufficient training represents another potential barrier for your employees. So, if our objective is to remove barriers, this one is no different and must be bulldozed along with the rest. Training must be important since Deming mentioned it in 3 of his 14 points. Correspondingly, Point 4 of our Model Vision states that we want our workforce to be "highly skilled professionals"; in this chapter, I attempt to define what that looks like in practice. I introduce the subject at this point in the discussion because it relates directly to Chapter 8 on the process control manual (PCM). Namely, you can put together the best PCM the world has ever known, and it will be of little value if your employees do not have the skills and training to put it to use.

Assessing Basic Skills

The first step in the process is to have your PCM professionally assessed regarding the minimum education level needed for a user to have full comprehension. After that, you will have to determine how well your employees' basic skill levels match those requirements.

We tend to take for granted that people who are capable of holding a job in a factory year after year have reasonably good language and math skills. However, from firsthand observation, there are a surprising number of people who do not. Also, the information in the PCM is not rudimentary. In fact, it is fairly sophisticated. So, if you expect to train people to use the manual the way it is intended, you must *know* that they have those essential skills. How can you know this? You will have to *test* them.

Getting employees to agree to basic skills testing can be a bit traumatic for some people because they are fully aware of their educational shortcomings. And for them it is embarrassing and maybe even frightening. So, it goes without saying that the testing must be confidential and nonthreatening. You must emphasize that the intention is to assess who needs additional training, not who will be demoted or shown the door. Explain to all employees that the PCM and statistical methods must become a way of life, and that the management team's job is to make each employee successful—period. While each person may have a different starting point in the learning process, your team is committed to doing whatever is necessary to help everyone get to the finish line. This may well include company paid "refresher training" in high school level English and math, which is available at nearly any local tech school. If there are no tech schools nearby, recruit local teachers to provide the training in your plant or in a local school after hours. Just do whatever it takes.

So, now let us assume the testing is completed, and you know the education level of your workforce. Bring *each employee* in for a private interview and share the results with the employee. Low-scoring employees must be enthusiastically encouraged to pursue the refresher training. Emphasize that they are valued employees, but that the nature of the work has changed. This new way of manufacturing is merely a response to global competition, and if nothing is done to meet that challenge, the plant and company would be at risk. The number one task is to stay in business. The employees are an integral part of that effort. However, there is one stipulation. Employees who refuse refresher training will have limited job opportunities in the future because they will not have the necessary skills. You are not demoting them or punishing them, but it will definitely limit the work they can be assigned in the future, while those who have the skills may get promotional opportunities that they could not otherwise been offered.

> Note: You can only afford to make this concession for a very limited number of employees. If the word gets out that all this is "optional," you can forget it. Few people will volunteer for the increased complexity without additional pay (which could actually be possible under the right circumstances).

For the high-scoring employees, you get the pleasure of congratulating them and showing your appreciation for their advanced skills. Let them know that you will be counting on them to help train others as the organization progresses to the new manufacturing system. There may even be opportunities for advancement if they are willing to help. But, do not say that if you do not mean it.

The Training Begins

As we said in Chapter 8, the PCM is the training "bible." *It must be ready before you start the training process.* You must train chapter by chapter in the order given. Do not jump to the part on statistical control and start control charting. Also, do not assume people know the process material. Yes, they know some of it or maybe even a lot of it. That is irrelevant because you are training on principles, and principles require a solid foundation.

You will need to introduce the subject of statistics separately and in sufficient depth that everyone understands the symbols and terms used in statistical process control (SPC) and how they relate to the process. Conducting the Red Bead Experiment (Appendix 1) in classes is strongly recommended to make the subject real. Again, you can purchase the kit (from http://www.qualitytng.com). Then, use Appendix 2 to introduce the standard SPC tools that they will be seeing in the future. Focus on the rules for recognizing when a process is "in statistical control" as introduced in Chapter 8. Then, move to control charts and the rules for when it is appropriate to make adjustments given in Appendix 2. Last, you will have to teach them to use the computer SPC program (like Statit®) you selected. This is a *lot* of information. It will take time and commitment from your entire team.

Training sessions are conducted on an overtime basis. The trainees attend on their normal work schedule and are replaced by others on overtime so the trainees are fresh. Coming in 2 hours early or staying late will not make them "fresh." You will also need a dedicated training facility and dedicated trainers. The trainers should be hourly employees who are the natural leaders in each department and are paid accordingly. Since nearly all of the material is technical in nature, your technical team trains the trainers and then functions as additional classroom resources. Your human resources team schedules and facilitates everything else. I would assume by now that you have surmised that none of this is inexpensive. As a really bright colleague of mine used to say, "Quality is free [after Philip Crosby's book of the same name], but it ain't cheap!" In other words, the initial investment is big, but once you have made it, elimination of defects, rework, and increased productivity pays for that investment. But I digress. The real payoff is you get to stay in business.

Verification of Training

No training is complete without testing to verify that the training objectives have been accomplished. Here again, people do not like tests, and some may complain. It is my personal belief that those who complain do so because they do not know the material or think that they cannot learn it. Remember that there is no room in this process for fear. Trainers and human resources must provide continual encouragement to ensure people are not afraid of the training or testing. While test results require absolute confidentiality, employees who score low will require additional training in small-group or one-on-one settings. Be prepared to pull out all the stops to show your commitment.

When your team has exhausted to no avail every measure practical to help an employee pass a test, work assignments for that employee must then be restricted to tasks that he or she can do successfully. There is no flexibility in this regard. The challenge is for you to do this in a way that is not obvious to the rest of the employees. In any case, the one thing you must *never* do is place an employee in a job for which the employee is not qualified. This can be costly for the enterprise. It is also unfair on many levels, and under the right circumstances can even be dangerous.

If you were interviewing a candidate for a position in your new model organization, you would have different selection criteria than you had in the past. However, your current employees were qualified under the old model and have invested their lives with your company. If they have been satisfactory until now, it is incumbent on you and your team to make them successful—period. Anything less signals that you are ruthless and uncaring, which will poison your organization like nothing else.

Retraining

Let us assume now that you have completed this massive training and testing program and your people have started SPC. If you have made it this far, congratulations. However, do not rest on your training laurels because you are not done yet by a long shot. After a few months, you need to go back and start over again, not from the very beginning, but say somewhere in the middle. By this time, two things will have happened: People will have been able to put what they learned in class to use, and they will find out what they did not learn and have solid questions to ask at follow-up training sessions. They will know what they know and what they do not know.

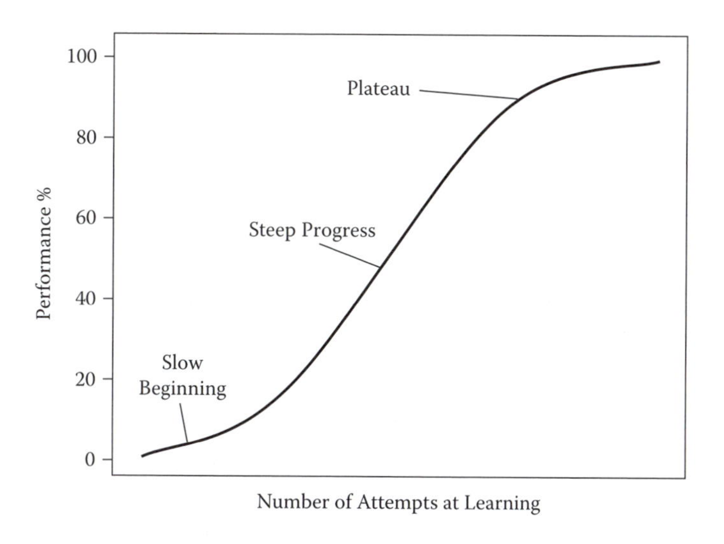

Figure 9.1 Typical learning S-curve

To get ready for the next sessions, you need to develop a comprehensive exam to see how much your people retained from the initial training. The test results will probably show you that, overall, people have retained about half of the original information. You then design new training materials to focus on the weak areas common to the most people. For individuals who have one or two specific weaknesses, you deal with them separately. Anyway, the point is you retrain your employees where they need it. You will hold their attention, and they will not resent it nearly as much. After the training, you test again, and it is hoped many of the gaps have been filled.

The basic deal is that training retention is more or less an asymptotic function. You can never get to 100% knowledge transfer for an entire group. You can approach it, but that is about it. Figure 9.1 is a typical "S curve" that illustrates the concept. It shows that you can train your people more and more, but you will get incrementally less and less. After only two or three repeats, it becomes not worth the effort. On the other hand, some of the gaps will fill themselves in over time with experience on the job. There are also "experience curves" to illustrate that, but we are getting a bit too far down in the hole.

Annual Recertification

One last step in this process is periodic recertification. Aircraft pilots are required to fly with a certified instructor once per year to prove they are still

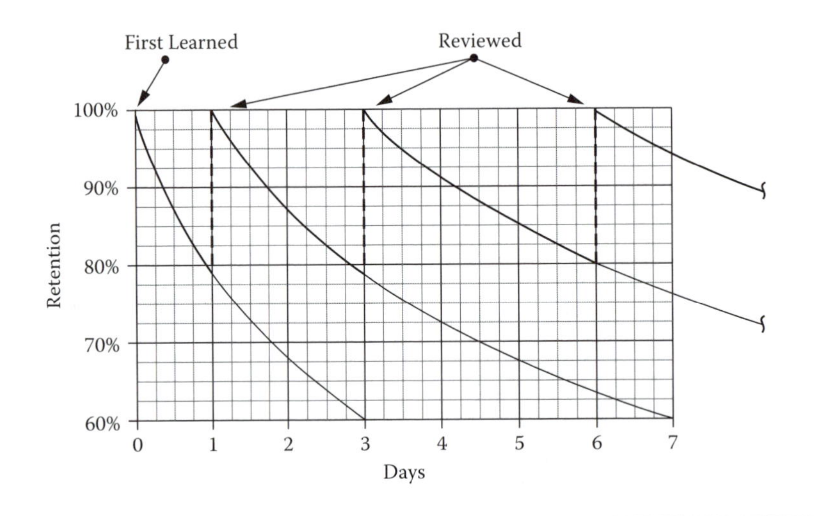

Figure 9.2 The forgetting curve. (Courtesy of Wikimedia Commons.)

qualified to occupy airspace. This applies to all pilots, not just commercial ones. The same principles apply to your employees. You need to be sure that they stay up to speed in process knowledge and procedures. People forget details with time (and age). Dr. Hermann Ebbinghaus developed a curve for this in 1885 called the "forgetting curve" (Figure 9.2). If this is accurate, people forget 40% of what they initially learned after only 3 days. It also illustrates that what they retain in subsequent retraining is remembered successively longer. So, if this is true and people forget that much, what do you do about it? You keep training. You will need to develop a refresher course that covers all the crucial details that are known areas people tend to forget. Administer the training annually and then retest and recertify—just like the pilots.

Summary

Deming has a lot to say about training. He often talks about people who are unable to do their jobs with pride in spite of "doing their best." You can do something about that by investing in the best training possible. You will be rewarded with a workforce who can take pride in their work and the company they work for. They will know you believe in them and are investing in the enterprise for the future—their future.

Chapter 10

Selected Topics on Working with People

Introduction

There is a book listed on Amazon.com entitled *If It Wasn't for the People … This Job Would Be Fun …* . The tongue-in-cheek implication is that managing *things* will almost always be more pleasant than managing people. Wishful thinking aside, the reality is you need quality people to operate and maintain your plant regardless of how automated it is. With that in mind, the purpose of this chapter is to present some selected topics that may make the job of managing people a little easier … and more fun.

Just like companies that say that safety is their number one priority, many of them also say, "People are our most important asset." But, if that is true, why is it that so many of the people I have worked with over the years did not feel that way about their company? Oh, do not misunderstand. These people greatly valued their high-paying jobs, but "managers" in general were not held in very high regard. And, in informal conversations, someone might wistfully recall fond memories of a past manager, but he or she always seemed to be the exception. So, what was the problem? Were those other managers' barriers?

From a personal standpoint, when I think about the managers for whom I have worked, only a very few made me feel valued as an employee. Almost none of them gave me counsel as a more experienced colleague might. It felt more like they were judging me on a daily basis. Instead of looking for something on which to compliment me, they were only interested in pointing out something I did not do or needed to do. In short, only a very few let me know that they knew what my contributions were and that they valued my work. And for those guys, I would have walked barefoot on burning hot coals. Anyway, from my personal experiences I am now convinced that this is the starting point for working with people—affirming their work, being appreciative, and removing barriers that hinder their success. From my view, there is absolutely *nothing* as effective or *rewarding* as positive reinforcement. As Larry the Cable Guy might say, "That's the truth right there. I don't care who you are."

Topic 1: Driving out Fear

We begin with the most important topic of all: driving out fear (Deming's Point 8). I used to love to quip that there are only two motivators in this world—greed and fear; but of the two, I preferred greed. Some people are put off by that, but I still feel it is mostly true in the workplace. If someone is motivated to succeed for personal advancement or monetary gain, I have fun calling that "greed." On the other hand, if someone else is only motivated to do his or her job because otherwise he or she will be fired, I say that is "fear." If you are an overbearing boss who gives little credit to others, you will get nothing from people in return but the latter. Fear is a terrible barrier.

Deming told stories of factory workers he encountered in his travels who were afraid to discuss their actual problems with their management because they literally were afraid of losing their jobs or the factory closing. What is unfortunate is that the information they held back due to fear may have helped keep the factory open.

I managed a small chemical plant for a few years, and I will never forget one of the employees who always prefaced his suggestions with, "I know you don't want to hear this, but … ." Boy, was he ever wrong. Whenever he said that, I knew he had some inside information about a nagging problem that had been neglected for years, but in the past, "No one would ever listen." Every one of his suggestions was pure gold, but he initially assumed I was like his previous boss who did not want to hear them. Not surprisingly, the more things I got fixed, the more information he would give me, and the more information he would give me, and so on.

Here is the bottom line: Fear is deadly. It takes all the joy out of work and can be a giant barrier between you and your employees, some of whom would love to offer their ideas and suggestions for improvements. Deming tells us that management cannot motivate people because, by nature, they are already motivated. Management, however, can surely demotivate them through fear or continued inaction.

Here is why: When a person offers unsolicited information, he or she is taking a risk that the person receiving the information might react in a way that makes him or her feel unimportant or irrelevant. These individuals will not do that more than once if the experience is negative. An effective manager will show appreciation and investigate promptly to see if an employee's suggestion is beneficial. If it is, it is a big deal. So, act like it. But whether beneficial or not, it is imperative that you follow up with a sit-down meeting to show the employee what you have discovered in your investigation. If you conclude that his or her idea is not helpful, you must be absolutely clear in presenting the facts because the employee will have his or her defenses up at maximum level. Regardless of merit, you must be appreciative of the effort and encourage continued input. This is your one chance not to lose an employee's participation, perhaps for life. Do not treat it lightly.

By the way, these principles do not just apply to top management. They must be driven into your entire organization if you expect people to work together as a team. Team members at all levels must be taught to be open to ideas from others. If not, people clam up and withdraw into cliques where they finally find sympathetic listeners, which eventually evolves into "b… [expletive] sessions." You can avoid this. There are some simple and practical team-building training materials out there that do not involve ropes or making people fall backward to be caught by somebody they currently dislike. Instead, the training demonstrates that a group in which everyone is participating is considerably stronger than individuals acting in silos. Making the entire team successful is as important as or more important than the success of an individual.

Finally, to have an organization that is free from fear, management must have the highest levels of integrity on the planet. Managers must have open doors and open minds and be perceived as caring and benevolent. Just like your children, the benefit of the doubt must always go in the employee's favor. Remember that you must reinforce repeatedly that management is an advocate. Employees go to management to get a problem solved.

Topic 2: Promotions (No Good Deed Goes Unpunished)

Promotions of hourly personnel into supervision can be a blessing or a curse. If the right person with the right qualifications receives a promotion, the entire organization benefits. Employees see this as a competent decision by their management. There is rejoicing. Life is good.

When management promotes the wrong person, the organization takes a hit. What are the circumstances that would cause employee dissatisfaction with a promotion? The person selected for promotion

1. Is not one of the established "informal leaders" in the workforce
2. Has less seniority than someone else with equal qualifications
3. Has undesirable personality traits
4. Has insufficient technical skills
5. Was selected as a result of successful "brownnosing"

The result will be that team members will not help him or her be successful. They will do just enough to get by or worse—they will do exactly as they are told, which means—they will not be a team.

Invariably, management will have a favorite candidate when it comes to picking an hourly employee for promotion. That person will often be a natural leader—a person who is technically competent and has personal confidence but is always looking to help others. He or she will be open to ideas by management and when given an assignment or a new project, will make every effort to make it work, including feedback on problem areas and suggestions for how to overcome them. Who wouldn't love this person?

Unfortunately, the superstar employee may be junior to others who have similar skills but are not as well rounded. The dilemma for management is to try to determine which promotion will have the best overall outcome because promoting the superstar may give one person a boost but demoralize several others. Those others may also be solid performers, and you cannot afford to lose them by being perceived as unfair or—worse—ungrateful for their years of dedicated service.

The Selection Process

Your only hope to get through this minefield unscathed is to have a selection process that is as objective as possible. What seems to work is a combination of a thorough technical exam, based on the process control manual, and a Targeted Selection® (TS) interview to *measure* behaviors.

The technical exam should be as comprehensive as possible and be in written essay form to help evaluate both written and technical skills. The exam is graded by a team consisting of one or more engineers and department management. Each team member performs a separate review of the examination and assigns his or her initial grade. The team then discusses individual assessments and jointly generates a list of questions on any exam area that is inaccurate or less than complete. The team then meets with each candidate to discuss its questions to see if any new information can be discovered. After all of this, the review team assigns a final grade.

On the behavioral side, TS is a technique used primarily for interviewing candidates for employment, but the principles apply equally well when selecting employees for internal promotion. It focuses on what a person has actually done in the past in certain situations that would predict what he or she would do in similar circumstances in the future.

The TS interview is conducted by a minimum of three trained interviewers who have diverse backgrounds. One of the interviewers should always be a human resources (HR) official, another must be the stakeholder, and the last an experienced colleague. There are typically five or six discussion questions that are designed to cover "targeted" behavioral characteristics like initiative, personal relationships, integrity, and ability to resolve conflict—areas in which the candidate will need to be successful in the new position.

For example, typical interview questions on initiative are something like the following:

Tell me about a time when you did more than was required on your job.

- What was the specific situation? (When, which job, which company, etc.)
- What did you do that exceeded expectations?
- Why did you do it?
- What was the outcome?
- What did your boss say?
- Did you get any recognition for it? If you did, what was it?

The interviewer must take accurate notes and stay focused on capturing actual behaviors. Nothing else will do. Any time a person begins to respond with, "Well, in that situation, I would have … ," the interviewer must immediately redirect the candidate to something that *the candidate personally did* in the past. You want names, dates, places. It has to be rapid enough that the candidate cannot "make it up." Based on the quality of the answers, a numerical score of 0 to 5 is assigned, and a score of 3 is considered acceptable. (Note: Before making a final decision, an effort should be made to verify the answers given by the finalist.)

Each interview is conducted in a private room with each interviewer asking all of the questions. Approximately 5 minutes are allotted for each question. During the interviews, the candidates are asked to provide different example situations from the ones given to previous interviewers if possible. Immediately after all the candidates have run the gauntlet, the interviewers get together and fill in a matrix to score each candidate on each question. Each score is discussed and justified by the interviewer. Very high and low scores must receive particular scrutiny and group consensus.

So, now you have some very objective technical *and* behavioral data on which to base your promotional decision. In general, someone who has done well or poorly during TS will know it, which is a huge benefit. So, if there are two technically capable candidates and one did not do well in TS, he or she should have less resentment toward the person who was selected and the management team as well. It is hoped the best person will have the opportunity, and the organization will be pleased with the outcome.

After the Promotion

Once a person is promoted to an increased level of responsibility, the management team must kick into high gear to get that person up to speed and ensure his or her success. The old sink-or-swim philosophy is a recipe for failure, and this is a critical period. If a person has no prior experience in supervision, the person may not know what the real job is. So, it is up to you to make sure that the person does. The critical first step is to provide a *written* job position description that outlines clearly what the assigned responsibilities are. When you give the person this document, have a discussion about each point so that it is clear what the management team expects. You also need to expound on your own personal philosophy and "style" and emphasize that the person's job is to be a servant/leader as opposed to a boss.

You must provide both formal off-site training and on-the-job coaching to help the person make the transition smoothly. Always encourage the individual to seek guidance from his or her manager when there is doubt about how to handle a particular situation. When you are out and about, stop by the person's office for informal chats. Look for something to compliment the individual on, like "The department looks good. Thanks," or "Someone stopped by my office the other day and told me they were glad we picked you." Do not be stupid about it. Just let the person know that you are pleased that he or she was selected, you are confident the person will do well, and that the individual has your complete support.

That is about it for promotions. This is a sensitive, difficult area because if you make a bad selection, it is *very difficult* to undo. Demoting someone to the ranks is traumatic for everyone. You have to get it right every time.

Topic 3: Pay Systems

Since we all work for one reason—to be paid—it should be obvious that how you pay people can be a positive or negative influence in their job performance. In the days of "piecework" pay, the employees with the highest manual skill and dexterity made the most money. Regardless, pay rates were draconian, and quality usually suffered. So, while it was theoretically *possible* to make some extra pay, the work was incredibly grueling. I think this is the reason you do not see many workers smiling in old factory photos. I had a summer job like that once. It was quite a revelation (and motivator). By the way, a form of piecework is still used today in telemarketing and customer service, like in the movie *Outsourced.* These jobs still do not pay very much.

In today's factories, machines do almost all of the work, so pay based on piecework is probably not too relevant. However, some companies do pay bonuses for production levels beyond a certain quota. Deming told us in Point 11 that even that is not a good practice because if more production is needed, it is *management's job* to provide a process capable of producing *quality* product in whatever quantities needed. It is the employee's job to oversee the process and report problems to management. When you attempt to make employees responsible for increasing production quantity or bribe them with money to make more product than the process is capable of, you get bad quality, injuries, and breakdowns. So, let us at least agree that pay systems that are based on output *quantity* are not helpful.

Without piecework, you are left with only two pay options: paying for skill level associated with each job *position* or paying everyone the same. We look at each one.

Let us assume a hypothetical production line or continuous process that has several operating stations that require a human attendant. Each position may require different skill and experience levels. This was certainly what I saw in the chemical and paper industries, where the control room operator had, by far, the position with the highest skill level. So, were each of these positions paid differently? Again, it depends on the company's philosophy. One company chose to pay according to position skill level, and another chose to pay each position the same.

Pay According to Job Position

Again, the assumption is we have a process that has several different workstations. They could be assembly points, steps in the process that require human intervention to load materials, or just taking samples and monitoring for deviations. From necessity, we will need to assign one job position the responsibility for maintaining overall control of the process. Logically, the person in the control position must be intimately familiar with *every* aspect of the process to be able to make adjustments that are appropriate for the observed conditions. Clearly, this last position will be the highest skill level. So, one choice is to evaluate each position for skill and physical effort and pay each one accordingly. Using this approach, positions are fixed (usually by seniority), and you only have to train each person to work the position above to cover vacancies. The advantage is you save money on training and may have reduced overall payroll costs. Employees who are uncomfortable progressing to higher skill levels are sometimes allowed the option to "freeze" at lower levels, and junior employees can progress around them. Permanent promotions are only possible through attrition, so an employee can end up working the same job for a number of years.

Single Pay Rate

In the approach using a single pay rate, each operator is trained to work *all of the jobs* in the process and rotates to a different position each week. Since all receive the same training, all receive the same pay once they are tested and certified on each position. Aggregate pay rates do not have to be significantly different from the pay-by-position scenario, but they could be. While there will be some increases in training costs, every operator understands the entire process, so you should have a higher-quality workforce. Another advantage is reduced boredom due to the job rotations. Organizations that use this approach stipulate in the hiring process that, to be considered for "permanent" employment status, employees must be capable of qualifying on all the jobs in a process area. In my opinion, this is by far the preferred pay system and is much more closely aligned with our Model Vision. The one drawback is that it can be difficult to retrofit.

Motivating Employees with Pay (Not)

Can you cause an employee to work at a higher overall level of performance by giving the employee a pay raise? This is not usually the case. However, there are a couple of conditions related to money that *may* cause an employee to "work harder."

The first is to threaten his or her security (classic fear): "If you don't get such and such done by the end of the day, you're fired." Since this would mean a loss of *all* pay, you are technically motivating the employee with money.

The second is to provide an opportunity to make a step change in income level that is so large that it would change the quality of life for the employee or the employee's family (good old-fashioned greed). I think there are possibilities for this in upper management positions that offer outrageous bonuses. It causes some of these people to live at the office and never go home. Unfortunately, there is usually no one at home when they eventually get there.

For the rest of us working stiffs, you cannot motivate us with money; you can only demotivate us. We come to work hoping to do our best every day because we enjoy taking pride in our work. If you pay us fairly, we are grateful for our jobs and content to do everything we can to please you. We were already doing our best when you gave us a raise, and now we are just happy that *you appreciate us*. When you do not appreciate us or do not pay us fairly, we tend to do less of the extras. We will still do our best because that is built in to our character, but the creativity that comes from loving our jobs may be stifled.

I have never met anyone who claims to have seen an hourly workforce perform at a higher level after an annual pay increase was announced. When you are paying people as a group and there is nothing to differentiate between individuals, the meaning of an annual pay raise tends to lose significance. On the other hand, I have seen a number of salaried employees' face light up when I gave them the biggest raise I could because they had *earned it*. It "closed the deal" on our relationship because they *knew without a doubt* that I appreciated their contribution and had championed their performance to higher management. I am convinced it was not the actual dollars that were significant because the difference between average and above average annual increases was only a couple of percent. It was the undeniable positive reinforcement.

Topic 4: Customer Service

According to Merriam-Webster, the word *customer* is defined as "one that purchases a commodity or service." Now I have yet to see a customer walk in and buy something in any of the plants where I have worked. I think that may have been because we had no showrooms, and all the products were produced in bulk. As far as I can tell, there is only one person in the supply chain who actually places the product you make in a "customer's" hand and collects the money, and it is not you or anyone in your plant. Yet, every person in your plant at every step in that supply chain has customers. So, who are they? More specifically, who is your customer? It is the people you personally supply with a product, processing step, or service.

Table 10.1 lists some of the relationships. In this table, the term *downstream* means the next processing step in a manufacturing process. In other words, each department is *accountable* to the department that receives its

Table 10.1 Manufacturing Plant Customer Service Relationships

Position	*Primary Customer*	*Secondary Customers*
Plant manager	Marketing/sales, department managers, plant employees	Other plant managers, community
Department manager	Downstream department manager, department supervisors and employees	Other department managers
Supervisor/crew leader	Downstream supervisor, crew members	Other supervisors
Crew members	Downstream crew members	Other department crew members
Maintenance	Assigned process area team members	Anyone who needs assistance
Engineering	Assigned process area team members	Other engineers
Technical support	Assigned process area team members	Other engineers

output. This includes the warehousing and shipping department. If there is a quality problem between two departments, the receiving department *is the customer*—period. While the main contacts in these relationships are members of supervision, hourly employees should be encouraged to visit departments or schedule face-to-face meetings to discuss problems and ways to improve. This is Deming's Point 9 and our Model Vision Point 10.

Notice that the higher the level in the organization, the more customers you have. In this table, it does not matter whether your customer represents a primary or secondary relationship; they are *all customers* and must be treated accordingly. What does that look like? Think of yourself as a salesperson whose livelihood depends on commissions. When you are trying to sell something to someone, it is natural to be on your best behavior. You act cheerful, friendly, *and helpful.* You answer any questions the potential buyer might have, and you point out all of the product's features and qualities. A truly honest salesperson will also acknowledge any shortcomings in the product the buyer may point out. In the manufacturing environment, you have to go a step beyond that and *volunteer* information about any known shortcomings in the product or service to your customer.

There is one group (as stated in Chapter 3) who seems to have more difficulty grasping the customer service concept than the others: maintenance. For whatever reason, these folks have a tendency to remain somewhat aloof from us lesser mortals. When your machine will not start, you call a maintenance tech, who, after disappearing into the bowels of the process, appears a few minutes later and says, "Try it now." You hit the button again, and this time the machine starts, and the tech promptly rides off into the sunset (or the break room) on a golf cart. While it may not come naturally, when maintenance crews adopt a true customer *service* attitude, good things begin to happen in your organization.

All of us have customers in our jobs. We are directly accountable to the people who receive the output of our work. If they are not satisfied, we have a problem that we need to resolve without mediation from a third party (that is the last thing we need). The same is true for anyone who comes to us for assistance. The person chose us for a reason—he or she thought we could help, and we are obligated to provide that help or direct the individual to the right person. I have been in plants where employees at every level acted like they were accountable to no one. They projected an attitude of "It's not my job. Why are you bothering me?" For whatever reason, I really do not think they took pride in their work. Unfortunately, a couple of those plants do not exist anymore, and they were quite large.

By the way, a comedian named Freddie Prinze had a signature line in the 1970s—"It's not my job, man"—that made him a celebrity. It would not have been funny if it were not so ubiquitous.

Topic 6: Uplifting Performance Reviews

How many performance reviews have you had in your life that were truly helpful? I am not sure I have had that many. Oh, a number of them were favorable, but whatever "constructive criticism" I got in a performance review was way too late in the process. In my opinion, a performance review should document what you, the manager, and your employee achieved *together* as the result of your leadership and coaching and the employee's contribution to that effort. Think of yourself as a football coach on the sidelines during the big game. You call the plays, and the employee attempts to execute them and make extra yards by adding improvements. When the employee has difficulties, you are there to remove barriers and coach the employee through it. As the employee gains experience and confidence, he or she requires less and less coaching and begins to add more and more of personal creativity to the job (if the employee is free of fear). That is when the employee begins to "exceed expectations."

While coaching is an important element, all of us need the equivalent of a playbook if we are to have any idea what it is we are supposed to accomplish. That playbook is the dreaded "personal objectives." Why are they dreaded? This is because we are often given objectives over which we have no control. So, we hate them and resent the people who force them on us. In a football game, if a lineman misses a block, the quarterback can be sacked. In the business world, sometimes people get sacked for not meeting objectives over which they had no control. The bottom line is that when you give a person objectives, you are giving them to yourself as well.

Coaches go back and review movies of the game with the team and try to figure out what they would do differently next time. A performance review is the same thing. And, in my opinion, it is as much an evaluation of what you as manager accomplished as your employee. If you did not get the results you wanted, I think you are about 50% responsible because you either failed to provide sufficient direction or did not remove enough barriers.

Performance reviews should be almost a celebration of accomplishments, goals achieved, or projects completed. There should be no surprises because throughout the review period you and your employee worked together as partners in a venture. At the end of the "season," it is natural to assess what needs to be done to score even more points next year. Here is an actual example of what this can look like:

> I was once assigned an employee who had a history of mediocre performance. To be honest, the position this person had as a regulatory consultant was not the most exciting and pretty much fell into the "thankless" category. No matter—that was the job, and it was a needed function. The challenge was to make it into something positive for the company and the employee. Because the job had provided no personal validation in the past, the employee had lost objectivity and had become defensive. It did not take too long to figure out that this had been caused by both insufficient direction and lack of support. The employee was technically competent but was being ignored.
>
> So, I did two things. I coached the employee on customer service techniques and provided solid backing from a distance. We had many little tag-up meetings at which I provided reassurance and suggestions for how to better satisfy our customers. Over time, we were able to celebrate some successes. Eventually, the employee overcame the past and went on to become an effective consultant. After a year or so, it was time for a performance review. During the review, we discussed how well the employee had met our objectives and what the areas of opportunity were. It was a favorable review, and at the end I was paid the highest compliment as a manager when the employee said that the performance review "was the icing on the cake." And all I did was review the facts.

Obviously, not all performance reviews are positive. Sometimes, a person is placed in a position that does not match his or her skill set, and you just cannot fix it. If the person has integrity and you have gone the extra mile to try to make him or her successful, *your first duty* is to try to reassign the person to a position that is a better match. This is especially true if *you* promoted someone to his or her level of incompetence (the Peter principle). Here is one example:

> I once had a production foreman who was completely "in over his head." He was a nice guy and a reliable employee, but he frequently made mistakes, and it became increasingly clear that I would have to make a change. Fortunately, someone brought to my attention that the foreman was a former truck mechanic, and it just so happened that we had an opening for one in the truck shop. When I met with him and told him we were reassigning him as a mechanic at his same pay rate, he was thrilled (and relieved). He was

very uncomfortable in his current position and was more than happy to be back in a job that matched his skill set. After the move, he did a fine job helping maintain our fleet of Peterbilts.

The one thing a manager cannot fix is a lack of personal integrity. Those types of performance reviews are not uplifting. It is a sad occasion when you have to inform someone that his or her shortcomings are so great that the person does not fit anywhere in your organization. For these situations, there is nothing you can do to fix it. So, in a way, I guess we are off the hook. But, for all other performance-related issues, I say managers are accountable to make their people successful—period.

Topic 7: Your Employees Do Not Have to Be Superstars

Deming loved to tell the story about the news headline that said something like, "Half of schoolchildren's reading scores are below average." Duh—well guess what? Not every one of your employees will be a superstar either. Regardless, you should be able to run all the plays your boss gives you in *your* personal objectives with a team of solid average performers. You will, however, have to provide more coaching than you would for a group of superstars who can anticipate situations and take aggressive countermeasures like Aaron Rodgers, Tom Brady, and Eli Manning. Be that as it may, effective coaching can enable "average" players to perform way above average.

When my son was in the sixth grade, he decided he wanted to play city league football. At this age, the main purpose of the league was to teach young boys the principles of football and teamwork and to have fun. Unfortunately, my son ended up on a team coached by a man who was there to win games for his own personal ego. It was no fun, and the experience was so negative that my son did not want to play the next year.
As the start of eighth grade approached, my son decided he wanted to try out for football again. He ended up on a team coached by an ex-college wide receiver who had been quite a star until he suffered a horrendous injury that left him nearly a quadriplegic in a motorized wheelchair. But, this coach still loved the game, and he loved teaching boys how to play it. He was assisted by a dad who was his arms and legs along with his 15-year-old son, who had the personality of a pit bull and reminded me of Kevin's older brother, Wayne, on the *Wonder Years* TV series. Under this remarkable coaching, the boys played their hearts out and finished the season with only one loss. As customary, we had our closing banquet at a pizza restaurant

> and handed out the trophies. In a quiet moment, my wife and I thanked the coach for making this season a joy for all of us. The coach thanked us for our support and confided that the team was intentionally composed of all the boys who were the last to be chosen. Those were the ones he *wanted*. I never forgot that lesson.

At one point, I was assigned responsibility to manage a team that provided a wide range of technical services for a large manufacturing facility. The team members were a diverse group of average folks who happened to be very dedicated to their jobs and the company. In spite of this, one of my colleagues remarked that I had inherited the "department of broken toys." But they were not. They were an outstanding group who had been in thankless positions for years. Their spirits may have been dampened, but they still had commitment and *integrity*. Once we set objectives *for ourselves* and committed ourselves to providing the best customer service possible, the team came to life. As before, I stayed in the background, removed barriers from a distance, and acted as personal consultant for individual team members. As the successes rolled in, the team grew in every way. I saw lots of smiles because we were having fun. That is because winning is fun.

Topic 8: Nonpunitive Discipline—The Last Resort

What do you do when one of your children misbehaves repeatedly? For some of us, it means it is time for the last resort—a spanking. It also means, as a parent, you have exhausted all other methods of trying to change an unwanted behavior. And, spankings only work on small children. Once children begin to grow up, spankings lead to bitterness and resentment. Similarly, grown adults in the workforce do not respond well to punishment like deprivation of pay, which can also lead to bitterness. So, if your objective is to make a poor performer into a better one, you may only get a bitter person going through the motions. In my opinion, punitive discipline is not effective.

If an employee has a recurring performance problem, the employee must be *given the choice* of continuing to perform poorly and lose his or her job or change the behavior and be restored to a normal working relationship. It cannot be emphasized enough that if the employee ends up being terminated, it will be *the employee's choice*—the employee refused to change. You will help any way you can and offer all of the company's HR help, but the choice is the employee's and the employee's alone.

The traditional stages of discipline used to attempt to change an employee's behavior are listed next. As you can see, they are all negative:

- Verbal reprimand (or "warning")
- Written reprimand (or warning)
- Unpaid suspension
- Termination

Here is a nonpunitive version:

- Coaching session with informal note to file
- Coaching session with offer of assistance and official note to file
- One day on paid leave for employee to develop action plan and contract
- Termination for breaking contract

Let us discuss each of the nonpunitive steps.

Coaching Session with Informal Note to File

In this initial meeting, you identify the unwanted behavior by opening the coaching session with something like the following:

- "I've noticed recently that you've left work early on several occasions."
- "You seem to be calling in sick a lot lately."
- "That was quite an argument you got into last week. That's happened several times recently."

Or, if it is job related:

- "Seems like you've had problems getting your product samples to lab on time recently."
- "I've been going over your paperwork, and there are a number on incomplete entries."

Each of these is a specific *behavior*. Never, ever tell someone he or she has a "bad attitude." For example, if you were to say, "I don't like your attitude" to someone like Clint Eastwood, he would likely say, "I don't blame you. I don't like it either."

You cannot measure attitude, but you can surely measure and quantify behaviors. You may not like doing it, but quantifying and documenting the number of times a person repeats the unwanted behavior, like being late or absent, is a critical component of the process.

So, in this first coaching session, you let the employee know that you have identified a behavior that has happened more than once, and you ask if there is anything you can do to help avoid this in the future. Be sincere in trying to determine if there is some barrier out there that is preventing the employee from getting back on track—but only a barrier over which management has some influence. If there are no barriers, and the employee says he or she will take care of it, then you have reached the first objective of getting the employee to *take ownership of the problem*. Of course, all this must be noted and put in a file but do not announce that you are doing this at this point. This first meeting must be kept very low key and nonthreatening. To summarize,

- The employee now knows that you know about the problem.
- You have offered to help.
- The employee has declined your help and assumed responsibility for fixing the problem.
- You place a note in the employee's file unannounced for future reference.

Second Coaching Session with Official Note to File

While there may have been some improvement in the employee's behavior after the initial coaching session, the change was temporary. The unwanted behavior has returned. You schedule another coaching session, and it begins with something like the following:

- "It looks like the problem you and I discussed a week ago has returned. You improved for a few days, but that's about it."
- "If I recall correctly, you said you'd get this resolved, and yet it happened again on [fill in the blank] separate occasions."
- "Is there something you'd like to share with me about what's causing this?"
- "Is there something I might be able to help you with?"

Your objective for this meeting is the same as before: to get the employee to own the problem. Once the employee gives you a firm commitment that it will not be repeated, you say, "Great! But understand that I'm going to hold you to it, and that I'm formally placing a record of this meeting in your personnel file." You write the note, sign and date it, and place it in front of the person to read. Some companies ask the employee to sign the note, but often, the employee will refuse.

One Day on Paid Leave for Employee to Develop Action Plan

The unwanted behavior returns. You have no choice but to move to the final stage. You hold what amounts to a "last chance" meeting. You must have an HR representative present. You open the meeting with the following:

- "Do you know why we're here today?"
- Wait for the employee's answer. A person of integrity will know and admit why. A person without integrity will play dumb.
- "I've asked a member of HR to attend to be sure you are represented."
- "After our last meeting, you again agreed to get your issue resolved, but unfortunately, that hasn't happened. On [fill in dates], you did it again. We've arrived at the point where I must inform you that the next time it happens you leave us no choice but to terminate you."
- "Tomorrow, I am giving you a day off *with pay* to write up an action plan that you must commit to implementing. The plan must be such that if you follow it, you will be able to avoid ________ [repeating the unwanted behavior in question] again."
- "The action plan will also serve as an employment contract between you and the company. When you return to work, you must report to me, and you and I will sign and date it with an HR representative present as a witness."
- "If you break the contract, you will sever your employment relationship, and we will be forced to terminate you."
- "This agreement and the conditions for your time off are confidential. No one will know why you are off unless you tell them. Simply inform your immediate supervisor that you have an emergency at home and need a day off."

When All Else Fails: Termination

If the worst happens, you will have given him or her every opportunity to change the behavior, and the employee made a choice. So, if you have to terminate that person, you can at least do it with a clear conscience. Regardless, it is still something all of us try to avoid because it affects not only one person but also the person's family. Senseless loss of a good-paying job is sickening.

Conclusion

I have seen several lives changed using nonpunitive discipline and the people restored to normal working relationships. On the other hand, I did not see the old "progressive discipline" approach do any good in a union environment. All it did was keep poorly performing employees on the edge of termination year after year while they played the system to their advantage. The system allowed them to stay that way because there were no real consequences.

> I remember one employee with a terrible attendance record. He reported to me after another of his many absences and said with a smile, "I guess you're gonna lay me off." I said, "No, I'm not. It doesn't do any good. All it does is cost the company overtime money and make your fellow employees have to work extra hours. Go back to work. But understand, the next time you miss, regardless of circumstances, you will be terminated." The shop steward came immediately to my office and said, "I can't believe you didn't lay that guy off." And I said, "It doesn't do any good. But understand that the next time he misses, I'll terminate him." The end of the story was that not too long after that, the employee in fact missed again, and we terminated him with barely a whimper from the union.

In contrast, nonpunitive discipline works. There is nothing more powerful than being sent home *with pay*. Also, it is not a public flogging like it is when a person is sent home for a week without pay. Done correctly, other employees should not know it happened. Yes, I know—there are no secrets in a plant. But at least your intentions are honorable, and the employee's family suffers no financial loss. All the employee has to do is decide to change—kind of like quitting smoking. Once you finally decide, you can begin the process in earnest. Until then, it is futile.

Closing Comments

For some of us, working with people does not come naturally; yet for others, it does. Regardless of what your starting point is, the intent of this chapter is to give you the benefit of some learned lessons that will make your transition to being a "people person" smoother and easier while avoiding some of the pitfalls that can be difficult to fix. The end result will be less stress on you and your employees and a more "motivated" workforce. Remember, management cannot motivate people, only demotivate them.

Chapter 11

Some Pointers on Working with Unions

While working with unions is usually more difficult than not, it can be done successfully if a relationship of mutual respect is established. The unions I have worked with respected one thing: competency. That means you are solving problems, removing barriers, and making the plant run efficiently. In return, the union expects not to be treated as an adversary (even though sometimes that is hard to do). My experience in working with unions is somewhat limited—one 6-year stint. But, during that time I had three different assignments, so I did notice a few things that you may find useful.

1. There are almost *always* two sides to a story, and an impartial, objective manager can usually resolve differences amicably. The presence of a union puts a third party in the way and makes resolution considerably more difficult. This is because the union works at convincing employees that if they did not have a union, management would not treat them fairly. Thus, without a shop steward, management would not listen to them or address their concerns. (If this is actually true at your facility, then you probably deserve a union.) As a result, management must work all that much harder to demonstrate every day that they can be trusted because every bad decision will be written down and used forevermore as examples why the union is needed.
2. Regardless of whether your plant has a unionized workforce, you must treat individual employees as if you did not. If someone comes to you with a suggestion to improve the process or operating procedures, you

must act on that request whenever possible. (Just make absolutely sure that it does not infringe on established work rules.) If you establish this behavior as the norm, employees and the union will see that management has priorities in the right order—removing barriers and preserving the enterprise. And, you will be seen as having integrity. No one can take anything away from that.

3. If you or your team makes a mistake interpreting the work rules and causes a loss of opportunity for overtime or extra pay for someone, in all likelihood the union will file a grievance. If you determine that it was management's fault, pay the grievance. Do not drag it out. The sooner it is settled the better. Eat your mistakes and learn from them to avoid repeating them and looking stupid. Your integrity is worth a lot more than a few hours of overtime pay.
4. Unions prefer to make all work assignment decisions based on seniority. This will be a huge barrier for your organization because the right person for a job is very often not the senior person. There will be considerable extra costs associated with this one aspect. Unions have one job classification: member. As far as they are concerned, all members are equally qualified, and the most senior member gets to pick the most desirable work assignments. The result of this is that you will be forced to use salaried employees to do things like training that you would prefer to be done by your hourly employees. Therefore, the management team in a union plant will have to have more people than a nonunion plant.

   Occasionally, the senior union member will be a satisfactory candidate for a trainer or some other special assignment. You will want to take advantage of that, but can you really afford to? Once you do, the union will take the position that this is "their work," and there will be a labor dispute if you ever revert back to doing the job with a salaried person.
5. Without a doubt, the best way to deal with unions is by being competent. If you remove the barriers that keep people from taking pride in their work, guess what? *They'll take pride in their work.* If your plant is clean and safe, operating smoothly, and the people are making lots of high-quality product, you should *not* have a union problem. The employees should be having too much fun.
6. Every time you make a work system change, expect the union to want the new job to have a higher rate of pay. You will have to have your facts together in order to persuade them otherwise. And, if there is any

doubt about the valuation of a job, you may have to make a *minor* concession for the union to "save face." This may be the most exasperating aspect of union relations. You redesign work systems to be more efficient or invest in more sophisticated process equipment, and in either case, the union wants you to pay more per hour for labor.

7. If your maintenance force is organized along strict craft lines, the costs are killing you. It is a huge barrier you must eliminate. How do you do this? This is accomplished by training your single-craft tradespeople to be multicraft. However, since this will result in a lot less overtime, do not expect them to make the transition at the same pay rates. From their standpoint, they will argue that they are adding more value and deserve a raise. So, in addition to the costs for the training and recertification, expect to negotiate an increase in pay rates. The whole deal will hinge on how realistic the new pay rates will be. Make sure you have a sharp pencil.

   Here is a real-life example from some years ago at a large union plant where maintenance was divided along strict craft lines:

   > A small, electromechanical latching device on our machines required three craftspeople to replace. It was held in place by two bolts, which were removed and replaced by a millwright. It contained a solenoid valve that had an electrical plug that had to be connected and reconnected by an instrumentation technician. Since it was hydraulically powered, it had hydraulic tubing connections, which were removed and replaced by a pipefitter. If the device failed in the middle of the night, we had to call in three different craftspeople and pay each of them the minimum of 2 hours of overtime for a total of 9 hours of straight-time pay. In reality, any of the shift mechanics could have replaced it in 30 minutes if allowed to do so. After the transition to multicraft technicians, that is exactly what happened.

   The new work rule must be that jobs can be done by anyone who has the training or ability to do the work. If you bargain properly, switching over to multicraft will pay big dividends in reduced maintenance labor costs and increased plant efficiency.

8. You cannot pay higher wages than the product can support. There is a clearly defined limit that if exceeded will cause your facility to lose its competitive position. The union needs to know what that is, and that excessive demands will put everyone's job at stake. There are too many cases in the news where management bargained away their rights or made wage concessions to avoid a strike. The result was that those

facilities were eventually closed. Do not expect the union to believe you cannot afford higher wages if all the managers are driving a BMW or a Mercedes-Benz or the company president got a $10 million dollar bonus.

9. If the cost of a strike is a threat to survival, then (get ready for this) management and the site's salaried personnel must be able to run the plant themselves at some reasonable production level. You may need assistance from employees at other plants, and you may even have to bunk on site in dormitories to get through it. But, if your choices are between an unaffordable wage and benefit concession or crippling strike outage, this is one way to be able to bargain from a position of equal strength. Your engineers and maintenance and productions managers will be the plant leaders. If you are the plant manager and you do not have their know-how, then you become just another worker bee, and they will love you for it.
10. If management is unable to bargain from a position of strength, the long-term result can be a slow death sentence for a facility or entire corporation. Over time, union work rules tend to become more and more restrictive. Pay and benefit demands become increasingly unrealistic, and the enterprise eventually slides into bankruptcy. If the owners have the will, a Chapter 11 reorganization has the potential to undo all the damage done by the years of incompetent bargaining. Unfortunately, as is often the case, the facility has been neglected due to the financial strain of the union, and there is nothing left to save.

In summary, if you do not have a union today, you do not want one. You should treat your hourly workforce with the appreciation they deserve for sticking with you by paying them as generously as possible. Removing the barriers that enable people to take pride in their work will help them appreciate you in return.

Chapter 12

Organizing for Success

Introduction

Manufacturing organizations have "personalities" based on the philosophy of the persons who set them up. There seems to be two distinct perspectives. For discussion purposes, I will call them the *interdisciplinary team* (IDT) concept and the *multidisciplinary team* (MDT) concept. Both concepts meet the definition of a team—a group of people working toward a common goal—but there are significant differences. Some might say that the IDT is modern, while the MDT could be considered "old school."

In the IDT, there is lots of overlap between disciplines, and boundaries are not clearly defined. People feel free to dabble in their "neighbor's" area of expertise, and often this is encouraged to get "fresh ideas." Everyone in the IDT reports directly to the manufacturing team manager and shares a common office area.

In the MDT, there is considerably less overlap between disciplines. Each team member is considered an expert in his or her field of discipline. While everyone is encouraged to share their ideas, they tend to restrict them to their area of expertise. Technical specialists may have a desk in the manufacturing team office area, but their real desk is back with their technical group, who they rely on for consultations and backup. And, while they are accountable to the manufacturing team manager for results, they officially report to their respective technical manager, who is a senior member of their discipline.

If you ask someone on the street what the job of manufacturing is, the person would likely say that it is to "make products" or words to that effect. While that may be true if you have a monopoly, the day you have to start competing, manufacturing becomes two jobs: making products *and improving the system.* The question is which team concept, IDT or MDT, does the best job of doing both?

The Interdisciplinary Team Concept

The IDT concept was supposed to be the wave of the future in manufacturing. Using this approach, the manufacturing team encompasses all supporting functions for streamlining and self-sufficiency. All of the technical disciplines, maintenance, and production resources officially report to one manager who is responsible for producing the product. As stated, overlap between positions and exchange of ideas is encouraged, which some people even refer to as "cross-pollination." While this seems to be a valid premise, in my opinion the IDT concept evolved due to dissatisfaction with the MDT. That was because support people working in functional silos were thought to be insulated from the consequences of poor manufacturing performance—unlike those in production, whose personal careers were on the line daily. To be more specific, engineers and maintenance technicians were not directly accountable to the production function and appeared to set their own priorities. So, the theory was if you gave them all the same boss, they would be considerably more responsive because they would be too afraid to do otherwise.

An example IDT organizational chart is shown in Figure 12.1. In it we see there are three product teams, with each containing all of the engineering and maintenance resources needed to sustain day-to-day operations. To the right are the process engineering and reliability leaders, who are there to provide mentoring to their respective technical resources in the product teams. In reality, both of these leadership positions may have insufficient influence since all of the real authority is vested in the product team managers. This has the potential for engineering resources to be relegated to putting out fires each day instead of being allowed the time to pursue analytical solutions that would improve the system.

The Multidisciplinary Team Concept

In contrast to the position overlap in the IDT, the MDT is more like a pro football team or a machine. Machines have individual parts, with each having a single function. For example, in your car's engine, the valves cannot

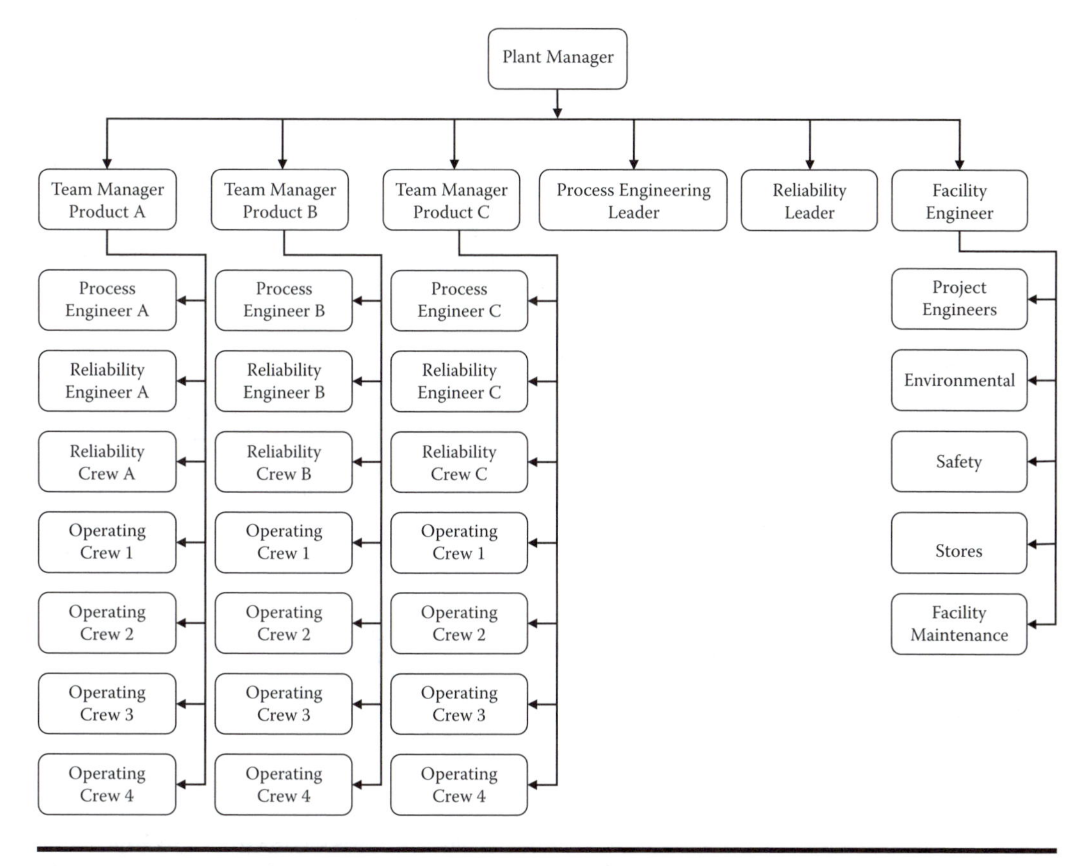

Figure 12.1 Manufacturing organization interdisciplinary team concept.

help out if one of the pistons starts leaking oil. But, when all the individual parts perform as designed, the engine runs smoothly. Similarly, on a football team, the interior linemen cannot help the wide receiver, who has to figure out how to get in the open by himself. And conversely, you do not see many wide receivers blocking up the middle. Just like an engine, each player has a different function, so different that each category must be coached separately to develop the skill levels high enough to be competitive. And yet, somehow, on game day they manage to all come together and execute complex plays with split-second timing. The only time players leave their playbook assignment and overlap with other positions is during a busted play. In my opinion, this is how the manufacturing team should operate. There should be no hidden agenda in the structure or need for one. Each team member is a professional and can be counted on regardless of reporting relationships.

Due to poor management in past years, individual manufacturing functions did in fact work in silos. If that were not true, Deming would not have admonished us in Point 9 to break down barriers between departments.

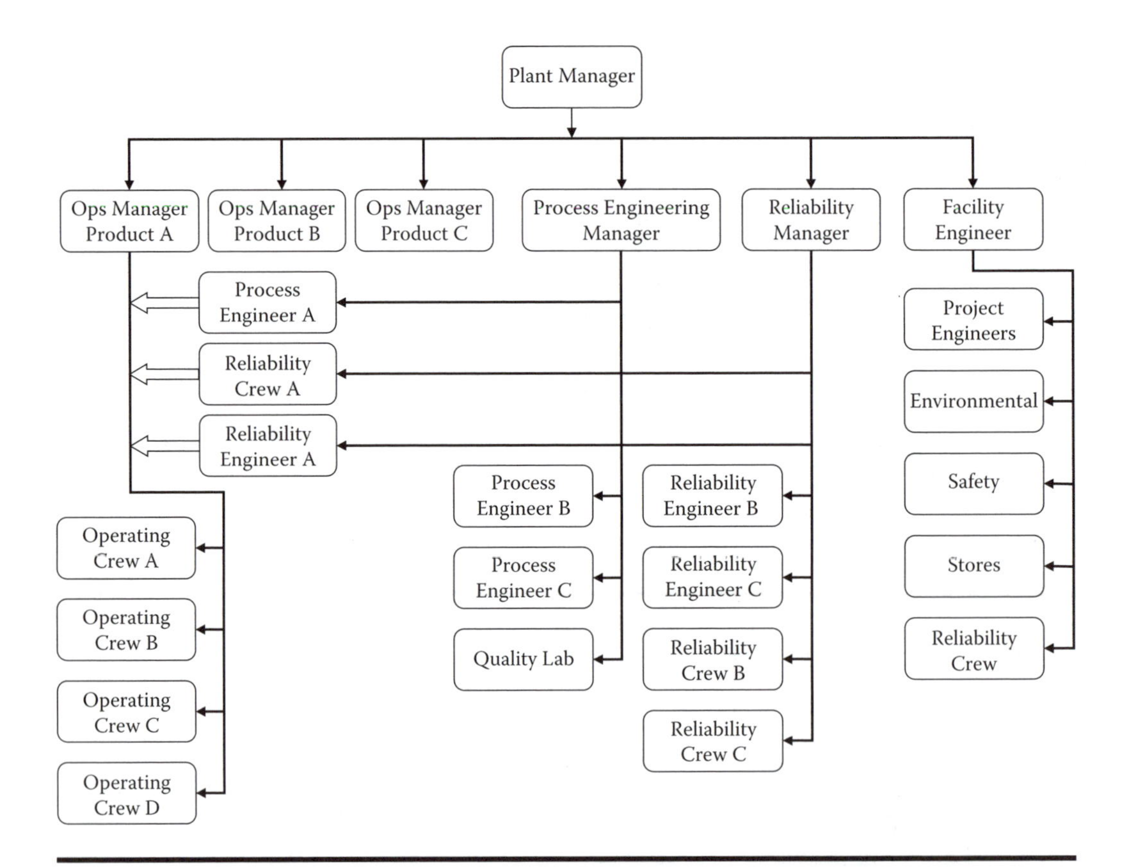

Figure 12.2 Manufacturing organization multidisciplinary team concept.

However, just like professional football positions need specialized coaching, manufacturing functions need separate, professional leadership. The "team concept" breaks down barriers between departments by eliminating the departments. I do not think that is what Point 9 indicates. I think it is more about customer service and establishing who the customers are. If the only way you can get people of different disciplines to work together is by having them report to the same manager, I think you have the wrong managers.

An example MDT organizational chart is shown in Figure 12.2. In this structure, the production function is led by an operations manager, but this designation is only to differentiate from the IDT for the sake of discussion. This position could still be called manufacturing team manager because the responsibility is the same: to make products. They have the same number and kind of resources working for them as in the IDT. The main difference is the technical functions do not officially report to them, so they must take advice and counsel from the technical and reliability leaders as full partners in the enterprise. Notice the oversize arrows from the technical support boxes to the product team. Those arrows have a name: *customer service*. The team's focus is twofold: making products and improving the system.

Now, it is time to look inside the boxes and see what each position actually does.

The Three Manufacturing Functions

In the past, one of the most brilliantly managed companies in the world specified only three manufacturing functions: production, technical, and mechanical. Some more modern names for these might be operations, process engineering, and reliability. Other companies may have more functions with fancier names and more overlap, but in my opinion it all distills to those three. We need to define how each of them should function to support our Model Vision.

Function 1: Operations

Operations Managers

The operations function is directed by managers who are professionals with a technical background. They were promoted into management because they were problem solvers who happened to have good organizational and people skills. They were also self-starters who required little in the way of supervision. Their personal integrity is without question, and they have spent their careers removing barriers for others. They understand the entire business at the plant from beginning to end and can communicate effectively to the people both above and below their position level. They know that the main thing is reducing variability, and they work every day to make it better. If you ask them, they can tell you what their top three sources of delay and waste are, what the trends are for each of them, and who is working on them. They have solid working relationships with the other managers—and especially with their customers.

Operations Crew Leaders

Operations crew leaders are former operators—always. They are selected using the process described in Chapter 10 because they have *both* superior technical skills and leadership abilities. As seasoned veterans, they know a lot about how things work inside and outside their department and can actually help the people assigned to their crew when they need assistance. They may be salaried, but they are paid overtime using the same rules as the people they supervise. They are the highest-paid member on the crew because their leadership is valued. They have solid relationships with crew leaders in other departments and make customer service a top priority.

Operators

Obviously, it is the operators that run the machines and make the product. To be effective, each operator on the team must be an expert (that term again) on the associated equipment and the product and has been trained and certified by an official board of "standardization operators." These people know the process control manual backward and forward and are *not* just "button pushers." Ideally, operators are also trained by the maintenance technicians assigned to their area to do some level of maintenance, like replacing consumable parts. They also have sufficient basic skills to assist the technicians when performing major maintenance and are routinely scheduled to do so during shutdowns. There is no real overlap in duties between operators and maintenance because operators do not do work beyond prescribed boundaries except perhaps in dire emergencies.

Function 2: Process Engineering

The process engineering group encompasses the "thinkers." They are the engineers and scientists who study processes and data and apply textbook engineered solutions to problems instead of using "the blunder method" (trying something in hopes that it will work and then trying something else when it does not). They rely mainly on scientific theory and statistical methods, so they routinely employ the tools given in Chapter 5 to do their analyses. They spend a lot of their time "testing the envelope" by running designed experiments on new materials and equipment to see if they can reduce variability or improve efficiencies. Their job is to make a good process better, cheaper, and faster.

Process Engineering Manager

Process engineering is managed by a topflight engineer or scientist, preferably with an advanced degree. His or her primary responsibility is to direct the cadre of process engineers, serve as their mentor, and ensure that each of them is working on the right things. This person must have in-depth process knowledge and advanced skills in statistical methods. When it comes to reducing variability, they are the "keepers of the flame." As a manager, this individual has to have some people skills, but face it, if he or she is a genius, the manager can get by with less than sterling diplomacy—but only if he or she is a genius. One skill any manager has to have is great customer service. So, genius or not, process engineering managers are paid a lot of money to reduce variability and make the process more robust. Without doing that, he or she is not being effective, and it is time to make a change.

Process Engineers

Process engineers are assigned to various manufacturing areas where they attempt to develop as much in-depth expertise as possible. If they are moved too often, they do not become the experts you need, and if they are not moved often enough, they become dissatisfied and dust off their résumés. In any case, a good process engineer is a self-starter who constantly looks for opportunities to make improvements. Process engineers need to be free of fear to try things and of being denigrated for trials that turn out to be unsuccessful. Customers of process engineers are their assigned operating teams, who give them business priorities. They have a big customer service arrow attached to their box for a reason. Their interaction with their operations team should be so seamless that it is hard to tell that they do not report to the operations manager. But, they do not. They *report* to the process engineering manager, *who manages their careers* (including continuing education) and helps them over humps or run major trials by providing backup process engineers when needed.

Quality Assurance

The quality function is a specialized branch of the process engineering group. It manages product sampling and analysis and the computerized statistical process control system. This group, which varies in size depending on the size and nature of the facility, also reports to the process engineering manager. It is composed primarily of quality technicians, who are usually former operators. This means they have a working knowledge of where specific defects originate in the process and what "good" product looks like. The group may have a full- or part-time leader assigned to coordinate activities.

In a Lean organization, it may be fashionable for operators to do their own quality testing. The main drawback to this is that it has the potential for increasing variability due to inconsistencies in the way people run the tests. Therefore, if you have multiple production lines making the same or similar product, it would seem to make more sense to have one person testing the output of all the lines to reduce "noise" in the test data. Of course, this has to be balanced against the labor costs and the amount of actual testing work that has to be done. However, if reducing variability is the main goal, a dedicated quality function is recommended.

Function 3: Reliability

Reliability is a relatively recent term that combines two functions from the past—maintenance and "plant engineering"—into one. Traditional maintenance consisted almost exclusively of fixing things that were broken. If something kept breaking, maintenance would call in a plant engineer or maintenance engineer to diagnose what was causing the recurring failures. These days, a different term is being used in place of "maintenance." It is now called *reliability*, which is more descriptive of what management actually wants from this function: *a reliable process*. Modern maintenance is now as sophisticated as any other discipline (see Chapter 3), and if yours is not, you are behind the times. There is simply no room for breakdowns in today's competitive marketplace.

Like it or not, maintenance/reliability is a legitimate manufacturing function, like operations and process engineering, which have individuals with nothing to do but twiddle their thumbs if the machines break down. As much as a third of a plant's operating budget may be spent on maintenance, yet it is sometimes given short shrift by management. In addition, process reliability has a huge impact on variability. So, why is maintenance given so little consideration? After years in the business, it is my considered opinion that it is because of decades of poor customer service, surly trade unions, high costs, and, yes, continued breakdowns. On a personal level, what is worse than taking your car into the dealer for service? The answer is not much.

Our Model Vision calls for our processes to be capable of running without breakdowns from one scheduled manufacturing run to the next—in fact, all the way to the next scheduled maintenance downtime. If your reliability team can do that, they are earning their pay. If they cannot, you are not doing something right, and you are wasting a *lot* of money.

In past years, maintenance has done such a poor job of customer service that the entire function has been downgraded to less than a full partnership. In fact, many companies have washed their hands of it and contract it out. In my opinion, you can do that with your buildings and grounds but not your process because process reliability is related directly to variability. In that regard, top-level management should want to know what the breakdown rate is for the company just like the level of waste. If they do not, and they are spending untold millions on maintenance, there has to be a reason. For the life of me, I cannot think of one.

Reliability Manager

The reliability manager is responsible for eliminating breakdowns—period. This person is a seasoned mechanical or electrical engineer with a broad range of experience. An advanced degree is preferable, but not necessary. He or she manages the manufacturing assets of the entire plant through the reliability teams and engineers. While the primary responsibility is elimination of breakdowns, he or she follows variability trends carefully to ensure reliability teams stay focused on the main thing: reducing variability problems that are related to process equipment. The reliability manager ensures that the entire team is service oriented and that the needs of the operating team customers are met.

The reliability manager is responsible for coaching and career development of the reliability engineers and provides them with direction and backup as needed to deal with major projects. He or she also provides advice and counsel for reliability crew leaders and deals with personnel issues.

Reliability Engineers

Reliability engineers used to be called "plant engineers" or "maintenance engineers." They are the mechanical and electrical engineers who keep the facility and the process equipment up to date on a long-term basis. They are routinely called in to analyze problems like recurring failures that the technicians have not been able to solve. They also do cost-benefit analyses on whether a piece of equipment is beyond economical repair. One of their primary tools is predictive maintenance (PdM), in which they are experts. Using engineering analysis, they spearhead the effort to eradicate breakdowns and equipment-related process interruptions. If you ask them what the breakdown rate is or the top three sources of delay in their area of responsibility, they can tell you.

Like process engineers, reliability engineers are assigned to specific operating teams, who set their priorities, and for that reason they have a big customer service arrow attached to their box. However, whereas process engineers may have almost daily involvement in the process, reliability engineers typically do not unless a specific problem has been identified or a trial is being run. Unfortunately, there is a tendency for operations management to want the reliability engineers involved on every breakdown, which is a misapplication of their talents. These people are engineers, not maintenance technicians.

> I want to digress a moment here and talk about the problem of insufficient pay for plant engineers in general. Considering their years of education and what it takes to prepare for this position, their pay is typically too low compared to that of a maintenance technician. In an effort to ameliorate that, some major companies actually pay engineers overtime under certain circumstances. In any case, many of the best engineers leave their vocation and go into management—just for the money. This can be a great loss of experience for a facility, especially if the engineer goes to a competing company. If your engineers are reducing your breakdown rate and variability, they are worth retaining.

Reliability Crew Leader

Reliability crew leaders manage crews of reliability technicians. They were promoted from the ranks because they were top-performing technicians with demonstrated leadership abilities—both of which are vitally important to be successful in this position. They were selected for promotion using the process described in Chapter 10. They may be salaried, but they are paid overtime using the same rules as the people they supervise. They are the highest-paid member on their crew because their leadership is valued. They are fully aware that reliability/maintenance is a service organization, so customer service is their highest priority after the safety of their crew members and plant equipment. As seasoned veterans, they know a lot about how things work inside and outside their department and can actually help the people assigned to their crew when they need assistance. They have solid partner relationships with crew leaders in other departments and respond without hesitation when called for assistance. They routinely consult reliability engineers in the decision-making process and provide assistance on engineering projects as needed.

Reliability Technicians

From a qualifications standpoint, reliability technicians should be school-trained professionals. In addition to a tech school associate's degree, they should receive intensive training by the manufacturers of the equipment they are being asked to service. Anything less will lead to repairs using the blunder method and will prove costly in the long run.

Technicians are grouped into reliability teams by manufacturing process area. They get their priorities from the operating teams they service, and *customer service* is their middle name. If there is not enough of them to handle a problem in their area, their crew leaders call in other plant teams and contractors to help. They are like the Marines: They get the job done, and they definitely take pride in their work.

Another responsibility of the reliability technicians is to train the operators they work with to do as many low-tech jobs on the machines as possible. The operators are then able to do these jobs themselves and assist the technicians on more difficult jobs that are beyond an operator's capability.

How the Three Functions Work Together

Operations is responsible for *operating* process equipment as designed within the parameters specified in the process control manual. Operations is also responsible for identifying conditions that do not conform to process control manual specifications or events that cause increases in product variability. Process engineers are accountable to *analyze* variability problems from a theoretical standpoint and recommend solutions. They must be constantly searching for ways to improve the process. Reliability engineers *redesign* or *replace* equipment to stop recurring failures or reduce variability based on process engineering or their own analyses. If there is one common point of accountability, it is not a person—it is the product. Their common goal is to produce a product that conforms to specifications with the least amount of variability possible.

If these three groups have to be told to work together, their managers are at fault. The customer relationship between them should be as compelling as the one between a salesperson on commission and a multimillion-dollar account. Each of the three groups should "go the extra mile" to assist the others without being told by anyone. Moreover, they should *want to* do so.

Performance Reviews by Customers

One way to ensure that people want to work together is by having the customers do the performance reviews of their "suppliers." This means that engineers and maintenance supervisors are formally reviewed by the operations managers of the areas they service. Yes, they also are reviewed by their functional manager, but their overall rating is a combination of the two, which has a bearing on salary increases and eligibility for promotions. So, while a professional receives coaching and career management from a functional leader, the professional's success or failure is largely determined by his or her customers. Thus, we have the best of both worlds: an organizational structure that promotes technical competence and customer service. In my opinion, there is nothing surreptitious in this approach. It is a reflection of how we are judged by our fellow humans each day for the services we render to one another.

We Still Need a Team

While the performance review is a technique that can drive accountability, it still does not ensure that the result will be a team. That can only come about through mutual respect based on individual ability and integrity. Just like a pro football team, a manufacturing team must be a group of *experts* working together toward a common goal. Team members are willing workers because they have not been demotivated by management and are *free of fear*. They all have personal integrity because the employees who did not have it have been weeded out. The reason they are all experts is because they have been *trained* by *experts* in their respective fields. Because of organization and clear direction, they all *know* what their individual responsibilities are, and they work together with precision. And, yes, they take pride in their work. They do not normally have much, if any, overlap except for emergency situations because they are a "Lean" organization and cannot afford the luxury of overlap. But in emergencies, they will all pull together and fill in for each other as needed because they believe in the enterprise and trust each other's abilities implicitly. With teams like this, management's job is to call the plays (provide direction), remove barriers, and get out of the way.

Chapter 13

Putting It All Together with Action Plans

Introduction

Everything we have done in this book has been to support our Model Vision, but to this point, it has been somewhat hypothetical. If you intend to actually put this to use, *specific actions* are needed. The easiest way to summarize everything covered by the preceding chapters is to provide action plans for each of the key positions. In other words, if you are in one of these roles, here is what you should actually be *doing* each day to reduce variability and improve the system. The action plans will be in the form of spreadsheets.

We are going to assume at this point that the plant is equipped with a computerized maintenance management system (CMMS), a computerized statistical process control (SPC) system, and a local-area network (LAN) with plenty of terminals.

For convenience, here is a relisting of our Model Vision:

1. Workplace is clean, well lit, properly ventilated, and free of safety hazards.
2. Machines are in tip-top mechanical condition and capable of running without breakdown from one maintenance downtime to the next.
3. The process is monitored and controlled with modern instrumentation.
4. Only high-quality raw materials are used.
5. The workforce is trained to be highly skilled professionals.
6. Operators monitor product characteristics using standardized methods and adjust settings using SPC techniques.

7. Product quality is measured and reported using appropriate statistical methods.
8. Each day, cross-functional teams analyze the process and product data and look for improvement opportunities.
9. Production schedules are easily maintained because surprise quality defects are nonexistent, and each operation is equipped to make changeovers rapidly.
10. Employees have the freedom from fear to try to make things better.

Action Plans by Position

Plant Manager

We start with the plant manager and work our way down in order of responsibility. As shown in Table 13.1, the plant manager is responsible for establishing expectations and maintaining them by monitoring the activities and results. He or she does this by walking around and making observations, providing funding for plant initiatives, reviewing reports, and holding periodic formal review meetings with the team.

He or she makes formal housekeeping inspections that cover the entire facility at least quarterly and performs a minimum of one announced inspection and at least one informal STOP® observation per month, which is the standard for every member of supervision. Formal monthly and quarterly reviews of the listed topics ensure that the team does not let any of the vision elements fall by the wayside while staying focused on the top issues.

Like all members of supervision, the plant manager has a LAN terminal in his or her office, which is scanned daily to keep informed on plant issues. In addition, there is a continuously updated plant summary report that shows all of the vital statistics on quality, productivity, and maintenance on one screen with hyperlinks to running trend charts. A sample report is shown in Table 13.2.

The plant manager works at driving out fear by being cordial to all employees and never goes out on the plant floor without smiling and greeting everyone he or she meets. Some plant managers pride themselves on being able to memorize the names of hundreds of employees. If you are able to do that, it is a nice touch. In any case, there is no excuse for not knowing the names of everyone in smaller plants. The plant manager is everyone's advocate and is never too busy to meet with an employee to try to help with a problem or listen to a new idea. If someone on the management team is having difficulty, he or she feels free to ask the plant manager for advice because he or she knows the plant manager will provide a sympathetic ear and wise counsel.

Operations Manager

Operations managers have very similar action plans as plant managers, but their "sphere of influence" is mainly confined to their operating department, as shown in Table 13.3. Other differences are the increased frequencies of their duties and level of detail. Good housekeeping and safety practices set the stage for success in other areas, so these are looked at daily. To ensure standard procedures are being used, this person should make one STOP observation per week.

The operations manager must be intimately familiar with equipment reliability issues in order to set priorities and direct team resources appropriately. While operations team members closely monitor SPC and process control systems, the operations manager is equally knowledgeable and scans the computer daily to keep abreast of any developing trends.

The operations manager provides the main leadership role for driving reduction of variability, which is a top priority for every department. He or she ensures that all process upsets are thoroughly investigated and prevented from recurring. In addition, there are standing teams that meet on a scheduled basis that continually work on the top three sources of delay and waste using the standard SPC tools.

The operations manager is the personal advocate for everyone in his or her department and mirrors the plant manager in style of management and professionalism.

Operations Crew Leader

Again, the action plan is scaled back to correspond to the scope of the position as shown in Table 13.4. Housekeeping and safety are monitored more frequently than once per day, and their direct supervision may be required to maintain standards. Crew leaders should make at least one STOP observation per month.

Because crew leaders have substantial process knowledge, they are frequently asked to provide input in problem solving or process improvement sessions.

Table 13.1 Plant Manager's Action Plan

Vision Element	*What*	*How*	*Who*	*When*
1	Workplace is clean, well lit, properly ventilated, and free of safety hazards.	Communicate housekeeping standards to entire facility, provide funding for housekeeping and safety projects and STOP training, perform formal housekeeping inspections and STOP observations, publish inspection scores by entire management team on official bulletin boards	Self, management team	Weekly/ monthly
2	Machines are in tip-top mechanical condition and capable of running without breakdown from one maintenance downtime to the next.	Provide funding and support for PM, PdM, CMMS, and TPM; hold recurring meetings to review implementation progress along with vibration trends, breakdown rates, and major equipment issues	Self, reliability manager, plant engineer	Monthly
3	Process is monitored and controlled with modern instrumentation.	Provide funding and support for LAN-based process control system, scan system periodically to stay informed on current issues	Self	Daily
4	Only high-quality raw materials are used.	Hold raw material meetings to review SPC documentation compliance and defect rates and overall satisfaction with current suppliers	Self, purchasing manager, process engineering manager, operations managers	Quarterly

5	Workforce is trained to be "highly skilled professionals."	Hold meetings to review training plans from each department, provide funding for employee continuing education, review assessment of overall employee education level	Self, operations managers, reliability manager, plant engineer, human resource manager	Quarterly
6	Operators monitor product characteristics using standardized methods and adjust settings using statistical process control techniques.	Provide funding to implement computerized SPC system, hold recurring meetings to review SPC progress and effect on variability	Self, operations managers, process engineering manager	Monthly
7	Product quality is measured and reported using appropriate statistical methods.	Hold recurring meetings to review ongoing quality concerns for each department, provide funding to resolve process issues, publish monthly quality report	Self, operations managers, process engineering manager	Monthly
8	Each day, cross-functional teams analyze the process and product data and look for improvement opportunities.	Provide funding for system improvement projects, hold review meetings to review top three variability reduction projects in each department, review monthly reports	Self, management team	Monthly
9	Production schedules are easily maintained because surprise quality defects are nonexistent, and each operation is equipped to make changeovers rapidly.	Provide funding to implement SMED, hold review meetings to report progress and current projects	Self, management team	Quarterly
10	Employees have the freedom from fear to try to make things better.	Maintain open door policy, employee advocacy, assistance, awards	Self, management team	Ongoing

Table 13.2 Plant Report

Safety	*Today*	*MTD*	*YTD*	*LTD*
Recordable incidents				
First aid cases				
STOP inspections performed				
Unsafe acts/unsafe conditions score				
Unsafe acts/unsafe conditions trend chart				
Environmental				
Leaks, spills, releases				
Stack nitrogen oxide (ppm)				
Effluent biological oxygen demand (BOD) (mg/l)				
Energy Efficiency (BTU/Units Produced)				
Energy trend chart				
Quality				
Lots held				
Composite sigma				
Composite sigma trend chart				
Sigma trend charts by department				
Composite percentage of target				
Composite percentage of target trend chart				
Percentage of target trend chart by department				
Productivity				
Composite production rate as percentage of standard				
Composite production rate as percentage of standard trend charts				
Production rate as percentage of standard by department				
Production rate as percentage of standard trend charts by department				

Table 13.2 *(Continued)* Plant Report

Safety	*Today*	*MTD*	*YTD*	*LTD*
Waste				
Composite percentage waste				
Percentage waste by department				
Percentage waste by department trend charts				
Top three waste source trend charts for entire plant				
Delay (Nonbreakdown Related)				
Composite percentage delay				
Percentage delay by department				
Percentage delay by department trend charts				
Top three delay items trend charts for entire plant				
Reliability				
Number of breakdowns				
Duration				
Breakdown trend charts				
Composite vibration level				
Composite vibration-level trend chart				
Current high-vibration equipment				
Number of air leaks waiting to be repaired				
Planned work orders completed				
Work order completion score				

Note: MTD = month to date; YTD = year to date; LTD = life to date.

Crew leaders are on the operating floor frequently, so they are aware of any ongoing problems with the process or the product. They know each of their crew members well and are aware of any training needs or personal problems that would be barriers to their success and work within the system to get them resolved.

The operations crew leader is the personal advocate for everyone in his or her crew and mirrors the plant manager in style of management and professionalism.

Table 13.3 Operations Manager's Action Plan

Vision Element	*What*	*How*	*Who*	*When*
1	Workplace is clean, well lit, properly ventilated, and free of safety hazards.	Perform daily walk-throughs, monthly formal housekeeping inspections, and STOP observations	Self, operations crew leaders	Daily/ monthly
2	Machines are in tip-top mechanical condition and capable of running without breakdown from one maintenance downtime to the next.	Implement TPM organization, review vibration trends and breakdown rates, set maintenance priorities	Self, reliability crew leader, reliability engineer, training coordinator	Weekly/ monthly
3	Process is monitored and controlled with modern instrumentation.	Become proficient in use of LAN-based process control system, scan system daily to stay current on plant conditions	Self	Daily
4	Only high-quality raw materials are used.	Review raw materials SPC documentation and defect rates, participate in plant manager quarterly reviews	Self, purchasing, process engineer	Monthly
5	Workforce is trained to be "highly skilled professionals."	Appoint training coordinator for department, ensure development and implementation of training plan, implement basic skills assessment and remediation, review status with team monthly and present to plant manager quarterly	Self, department training coordinator, operations crew leaders	Monthly/ quarterly

6	Operators monitor product characteristics using standardized methods and adjust settings using SPC techniques.	Maintain operator training and recertification, review SPC reports	Self, process engineer, process engineering manager	Monthly
7	Product quality is measured and reported using appropriate statistical methods.	Review SPC and monthly quality report, prepare quality presentation for plant manager review	Self, process engineers, operations crew leaders	Monthly
8	Each day, cross-functional teams analyze the process and product data and look for improvement opportunities.	Attend daily operations team meeting; establish standing teams to reduce variability, waste, and delay; support system improvement projects in budgeting process	Self, process and reliability engineers, operations crew leaders, operators, technicians	Daily
9	Production schedules are easily maintained because surprise quality defects are nonexistent, and each operation is equipped to make changeovers rapidly.	Monitor SPC reports, focus team on top quality and variability issues, establish and provide ongoing support for SMED team, observe and participate in development of SMED practices	Self, SMED coordinator, process engineers, operations crew leaders, reliability engineers	Daily/ weekly
10	Employees have the freedom from fear to try to make things better.	Open door policy, employee advocacy, assistance, awards	Self	Daily

Table 13.4 Operations Crew Leader's Action Plan

Vision Element	*What*	*How*	*Who*	*When*
1	Workplace is clean, well lit, properly ventilated, and free of safety hazards.	Make frequent walk-throughs and daily informal STOP observations, perform one formal STOP observation each month.	Self	Daily/ monthly
2	Machines are in tip-top mechanical condition and capable of running without breakdown from one maintenance downtime to the next.	Monitor process, consult with operators, report problems	Self	Daily
3	Process is monitored and controlled with modern instrumentation.	Scan LAN-based process control system	Self	Hourly
4	Only high-quality raw materials are used.	Review raw materials SPC documentation, report defects	Self, process engineer	Daily
5	Workforce is trained to be "highly skilled professionals."	Review crew training plan, schedule training as needed	Self, department training coordinator, operators	Monthly/ quarterly

6	Operators monitor product characteristics using standardized methods and adjust settings using SPC techniques.	Monitor SPC charts	Self, operators	Hourly
7	Product quality is measured and reported using appropriate statistical methods.	Monitor lab data	Self, operators	As needed
8	Each day, cross-functional teams analyze the process and product data and look for improvement opportunities.	Lead daily team meeting, report system anomalies, assist in setting priorities, participate in problem-solving sessions when needed	Self	Daily
9	Production schedules are easily maintained because surprise quality defects are nonexistent, and each operation is equipped to make changeovers rapidly.	Provide input to SMED team to improve procedures and equipment	Self, operators	Daily/ weekly
10	Employees have the freedom from fear to try to make things better.	Employee advocacy, assistance	Self	Daily

Process Operator

The process operator is responsible for knowing and following the procedures in the process control manual. He or she keeps assigned process areas clean and free of clutter. The main job is to operate the process within established parameters and make product that meets specifications. The associated action plan is given in Table 13.5.

The process operator has been thoroughly trained on all of the equipment in the department and can perform any of the associated tasks safely using the prescribed tools and safety equipment.

The process computer is the main input/output point and the process operator is an expert on all of the screens. Due to extensive SPC training, the operator knows when it is appropriate to make process setting adjustments and when not. He or she promptly reports process deviations from standards and determines causes of process upsets.

The process operator is free to make suggestions for system improvements and is sometimes asked to participate in problem-solving sessions. These operators are team players who are always ready to assist other operators when called to do so.

Process Engineering Manager

The process engineering manager's role is that of an expert consultant and mentor for technical professionals. The corresponding action plan is shown in Table 13.6. All members of management make plant walk-throughs and monthly STOP observations, and this position is no exception.

The process engineering manager owns the SPC system and is directly responsible for managing quality trends. Through his or her process engineers, the process engineering manager ensures that variability reduction is approached from a standpoint of rigorous math and science. This person also participates in raw material quality decision making.

As a mentor, the process engineering manager ensures that each process engineer has sufficient training and experience to tackle assigned problems and provides additional resources to cover large projects as a team. This department is fearless but not reckless. Experiments are rigorously designed to ensure reliable results. The process engineering manager ensures that engineers are recognized for their contributions and compensated accordingly.

Process Engineer

From an action standpoint, the process engineers are the people responsible for actually improving the process, which is reflected in Table 13.7. Like other members of management, process engineers are also required to make one STOP observation per month.

In the quest for reducing variability, process engineers monitor data for developing quality trends. They verify that control charts are being used properly, that process variables are in control, and that raw materials meet specifications.

Process engineers are masters of the SPC tools and frequently plan trials to investigate causes and effects in search of optimum operating points for increased productivity or enhanced quality. They are also routinely called on to lead or participate in team problem solving.

Quality Technicians

The quality technician is assumed to be working in a dedicated lab, which is the technician's area of responsibility for safety and housekeeping. See Table 13.8 for the corresponding action plan. The main role in reducing variability is running standardized tests as consistently as possible on both products and raw materials. This is a vital but boring job. The challenge is to be able to do this week after week and not lose focus. An additional duty is to monitor control charts to ensure that operators respond appropriately to the data they input to the SPC computer.

Reliability Manager

The reliability manager is responsible for (you guessed it) the reliability of the entire facility. The associated action plan is shown in Table 13.9. Like all members of management, this person is required to walk through the facility once per week and make at least one STOP observation per month.

The reliability manager is specifically responsible for eliminating breakdowns by implementing state-of-the-art maintenance systems and reengineering or replacing problem equipment. He or she does this through the reliability engineers and reliability crews, who must be well trained on the process, associated equipment, and maintenance systems if they are to be successful. Retaining and developing engineering talent is another priority.

Table 13.5 Process Operator's Action Plan

Vision Element		*How*	*Who*	*When*
1	Workplace is clean, well lit, properly ventilated, and free of safety hazards.	Follow prescribed operating procedures using specified safety equipment, perform cleanup duties, store tools and supplies in prescribed locations, tag and barricade hazards	Self	Daily
2	Machines are in tip-top mechanical condition and capable of running without breakdown from one maintenance downtime to the next.	Monitor process, report problems, suggest improvements	Self	Daily
3	Process is monitored and controlled with modern instrumentation.	Maintain process within specified limits using LAN-based process control system, make frequent walk-throughs of process, fill out process reading sheets per schedule, report problems	Self	Frequently
4	Only high-quality raw materials are used.	Report raw material defects	Self, process engineer	Daily
5	Workforce is trained to be "highly skilled professionals."	Attend scheduled training, request other training as needed	Self, operations crew leader	As needed

6	Operators monitor product characteristics using standardized methods and adjust settings using SPC techniques.	Monitor SPC charts, adjust process settings using SPC rules	Self	As needed
7	Product quality is measured and reported using appropriate statistical methods.	Monitor lab data on SPC computer	Self, quality lab	As needed
8	Each day, cross-functional teams analyze the process and product data and look for improvement opportunities.	Report anomalies, participate in problem-solving sessions when needed	Self, operations crew leader	As needed
9	Production schedules are easily maintained because surprise quality defects are nonexistent, and each operation is equipped to make changeovers rapidly.	Provide input to SMED team to improve procedures	Self	When discovered
10	Employees have the freedom from fear to try to make things better.	Make suggestions for system improvements, encourage other crew members	Self	When discovered

Table 13.6 Process Engineering Manager's Action Plan

Vision Element		*How*	*Who*	*When*
1	Workplace is clean, well lit, properly ventilated, and free of safety hazards.	Walk through each manufacturing department, make STOP observations	Self	Weekly/ monthly
2	Machines are in tip-top mechanical condition and capable of running without breakdown from one maintenance downtime to the next.	Monitor process and product variability for indications of equipment problems	Self, process engineers	Monthly
3	Process is monitored and controlled with modern instrumentation.	Scan LAN-based process control system	Self	Daily
4	Only high-quality raw materials are used.	Participate in raw material SPC issue meetings	Self, process engineers, operations manager, purchasing manager	As needed
5	Workforce is trained to be "highly skilled professionals."	Develop continuing education plan for process engineers	Self, process engineers	Yearly

6	Operators monitor product characteristics using standardized methods and adjust settings using SPC techniques.	Monitor variability data for potential overcontrol/undercontrol	Self, process engineers	Monthly
7	Product quality is measured and reported using appropriate statistical methods.	Audit lab processes, schedule periodic refresher training and recertification	Self, quality lab	Monthly/ yearly
8	Each day, cross-functional teams analyze the process and product data and look for improvement opportunities.	Verify improvement teams are functioning, provide expert consulting when needed	Self, operations teams	Monthly/as needed
9	Production schedules are easily maintained because surprise quality defects are nonexistent, and each operation is equipped to make changeovers rapidly.	Analyze defect reports, direct process engineering resources accordingly, verify effectiveness of SMED settings on quality of products after changeovers	Self	As needed
10	Employees have the freedom from fear to try to make things better.	Open door policy, employee career development, advocacy, project assistance	Self	As needed

Table 13.7 Process Engineer's Action Plan

Vision Element		*How*	*Who*	*When*
1	Workplace is clean, well lit, properly ventilated, and free of safety hazards.	Walk through assigned manufacturing department, make monthly STOP observations	Self	Daily/ monthly
2	Machines are in tip-top mechanical condition and capable of running without breakdown from one maintenance downtime to the next.	Monitor process and product variability data for indications of equipment problems	Self	Daily
3	Process is monitored and controlled with modern instrumentation.	Scan LAN-based process control system	Self	Daily
4	Only high-quality raw materials are used.	Investigate raw material issues and review raw material SPC documentation	Self, operations manager, operations crew leaders, operators	As needed
5	Workforce is trained to be "highly skilled professionals."	Develop expert-level process knowledge, request additional training as needed	Self, process engineering manager	Ongoing

6	Operators monitor product characteristics using standardized methods and adjust settings using SPC techniques.	Monitor variability data for potential overcontrol/undercontrol	Self	Daily
7	Product quality is measured and reported using appropriate statistical methods.	Monitor quality trends and SPC control charts	Self, quality lab	Daily/as needed
8	Each day, cross-functional teams analyze the process and product data and look for improvement opportunities.	Participate in problem-solving sessions and lead system improvement efforts, develop designed experiments to reduce variability and enhance quality	Self, operations teams	Ongoing
9	Production schedules are easily maintained because surprise quality defects are nonexistent, and each operation is equipped to make changeovers rapidly.	Analyze defect reports, provided technical assistance to SMED team as needed	Self	As needed
10	Employees have the freedom from fear to try to make things better.	Test new raw materials, determine process capability, verify optimum process settings, mentor junior engineers	Self	Ongoing

Table 13.8 Quality Technician's Action Plan

Vision Element		*How*	*Who*	*When*
1	Workplace is clean, well lit, properly ventilated, and free of safety hazards.	Maintain quality lab in clean and orderly condition at all times.	Self	Daily
2	Machines are in tip-top mechanical condition and capable of running without breakdown from one maintenance downtime to the next.	Ensure all lab equipment is in good operating condition, submit work requests for repairs as needed	Self	Daily
3	Process is monitored and controlled with modern instrumentation.	Verify all lab instrument calibrations are up to date	Self	Daily
4	Only high-quality raw materials are used.	Run prescribed raw material tests, report raw material issues to operations when detected	Self, operations crew leader, process engineer	As needed
5	Workforce is trained to be "highly skilled professionals."	Receive training and annual recertification on standardized testing methods	Self, process engineering manager	As needed

6	Operators monitor product characteristics using standardized methods and adjust settings using SPC techniques.	Check control charts in SPC computer to verify adjustment rules are being followed	Self	Ongoing
7	Product quality is measured and reported using appropriate statistical methods.	Perform quality testing per standardized methods, enter results in SPC computer	Self	Ongoing
8	Each day, cross-functional teams analyze the process and product data and look for improvement opportunities.	Provide data to improvement teams or perform additional testing as needed	Self, operations teams	Ongoing
9	Production schedules are easily maintained because surprise quality defects are nonexistent, and each operation is equipped to make changeovers rapidly.	Notify operating teams immediately of product defects or adverse trends	Self, operators, operations crew leader	As needed
10	Employees have the freedom from fear to try to make things better.	Make suggestions for reducing variability in test procedures or methods, assist other lab technicians and process operators	Self, other lab technicians, operators	Ongoing

Table 13.9 Reliability Manager's Action Plan

Vision Element		*How*	*Who*	*When*
1	Workplace is clean, well lit, properly ventilated, and free of safety hazards.	Walk through each manufacturing department, make informal job site STOP observations and one formal monthly STOP observation	Self	Weekly/ monthly
2	Machines are in tip-top mechanical condition and capable of running without breakdown from one maintenance downtime to the next.	Implement preventive (PM), predictive (PdM), and operator-assisted (TPM) maintenance systems, monitor PdM data and breakdown rates for adverse trends	Self, reliability engineers, reliability crews	Ongoing/ monthly
3	Process is monitored and controlled with modern instrumentation.	Implement process instrumentation maintenance and calibration systems, ensure process control issues are resolved promptly and permanently	Self, reliability engineers, reliability crews	Ongoing
4	Only high-quality raw materials are used.	Not applicable	Not applicable	Not applicable
5	Workforce is trained to be "highly skilled professionals."	Develop and implement training plan for reliability crews and continuing education plan for reliability engineers	Self, reliability engineers, reliability crews, training coordinator	Ongoing

6	Operators monitor product characteristics using standardized methods and adjust settings using SPC techniques.	Review overall variability trends for equipment-related opportunities	Self, process engineering manager, reliability engineers	Monthly
7	Product quality is measured and reported using appropriate statistical methods.	Not applicable	Not applicable	Not applicable
8	Each day, cross-functional teams analyze the process and product data and look for improvement opportunities.	Provide the highest levels of customer service, participate in improvement teams, provide expert consulting when needed	Self, reliability engineers, reliability crew leaders	Monthly/as needed
9	Production schedules are easily maintained because surprise quality defects are nonexistent, and each operation is equipped to make changeovers rapidly.	Participate in SMED teams to provide engineered solutions to reduce changeover times	Self, reliability engineers, reliability crew leaders	As needed
10	Employees have the freedom from fear to try to make things better.	Open door policy, employee career development, advocacy, project assistance	Self	As needed

Reliability (or maintenance) is a service function, and every team member is trained to provide the highest levels of professionalism. Reliability technicians are the highest hourly paid group in the facility, and it is incumbent on the reliability manager to ensure that his team adds commensurate value.

The reliability manager's management style is a reflection of the plant manager and the rest of the organization, which is based on effective coaching and employee advocacy.

Reliability Engineer

The reliability engineer provides the analytical branch of the reliability team. And, yes, he or she makes frequent walk-throughs of assigned plant areas and performs at least one STOP observation per month. The associated action plan is shown in Table 13.10.

The reliability engineer interfaces frequently with the reliability crews and serves as their technical consultant. This person is a master of the computerized PdM system and, along with the PdM technicians, follows trends closely to ensure breakdowns are minimized and maintenance strategies are effective, including replacement of obsolete or ineffective process equipment. For these and other reasons, the reliability engineer must receive whatever training necessary to qualify as an in-house expert.

The reliability engineer is often called on to participate in problem solving and SMED (single minute exchange of die) teams to apply personal expertise or develop an engineered solution to a needed system improvement. Because the organization is open to input from all participants, he or she feels free to contribute and make suggestions for system improvements.

Reliability Crew Leader

The reliability crew leader is responsible for actually getting the maintenance-related work done. As a member of management, this person is also responsible for performing one STOP observation per month. The position's action plan is shown in Table 13.11.

The reliability crew leader is an expert on the process and equipment within his or her area of responsibility and able to assist crew members with troubleshooting and problem solving. When needed, the crew leader calls in engineers or manufacturer's representatives for additional assistance.

From a customer service standpoint, the reliability crew leader attends the daily operations meeting to determine the day's priorities and assigns resources accordingly. The reliability leader also directs the crew's maintenance planner to ensure that there is sufficient planned work available to keep the crew fully employed every day and that schedule commitments are met.

Reliability Planner

From a safety standpoint, the reliability planner must ensure that all the safety equipment and lifting devices are inspected and tested annually and that all of this is incorporated into the CMMS.

The reliability (or maintenance) planner must be able to exploit the capabilities of the CMMS to maximum advantage by ensuring that all of the required preventive maintenance (PM) chores are completed on schedule while developing a library of as many preplanned jobs as possible. The corresponding action plan is shown in Table 13.12.

By the way, there are two schools of thought on maintenance planning. Some say that maintenance planning is an outdated function, and that technicians can plan their own work. However, in my opinion, it would be a very rare group of technicians indeed who had the personal discipline to mine all the capabilities from a CMMS, ensure that job plans were diligently closed out and updated using the Shewhart cycle, while creating equipment bills of material and keeping them up to date. In fact, a competent reliability planner can double the capability of a crew.

Reliability Technician

The reliability technician's job starts with a clean and orderly workplace, followed by performing his or her work using the appropriate tools, procedures, and safety equipment. The corresponding action plan is given in Table 13.13.

Technicians must be capable in all three maintenance systems: PM, PdM, and total productive maintenance (TPM). One of their main PM tasks is to complete work orders properly by providing the information needed to improve jobs the next time they are performed. While PdM is usually performed by a few specialists, all technicians should be familiar with the process. In the area of TPM, the technicians have to willingly "farm out" some of their work to operators. And, for this to work as intended, technicians must regard those operators as bona fide apprentices—*their apprentices.*

To enhance their value to the organization, technicians must place emphasis on customer service, just like the contractor workers do who show

Table 13.10 Reliability Engineer's Action Plan

Vision Element		*How*	*Who*	*When*
1	Workplace is clean, well lit, properly ventilated, and free of safety hazards.	Walk through assigned manufacturing departments, make informal job site STOP observations and one formal monthly STOP observation	Self	Weekly/ monthly
2	Machines are in tip-top mechanical condition and capable of running without breakdown from one maintenance downtime to the next.	Develop preventive (PM), predictive (PdM), and operator-assisted (TPM) maintenance systems, monitor PdM data and CMMS breakdown rates for adverse trends, engineer solutions to process bottlenecks, and eliminate recurring problems	Self, reliability crew	Monthly
3	Process is monitored and controlled with modern instrumentation.	Develop maintenance and calibration systems for process instrumentation and verify they are followed, develop hardware and software solutions to eliminate process control issues	Self, reliability crew	Daily
4	Only high-quality raw materials are used.	Not applicable	Not applicable	As needed
5	Workforce is trained to be "highly skilled professionals."	Develop expertise on assigned process and associated equipment, request specialized training from equipment manufacturers as needed	Self, reliability manager, equipment manufacturers	Yearly

6	Operators monitor product characteristics using standardized methods and adjust settings using SPC techniques.	Review overall variability trends for equipment-related opportunities	Self, process engineers	Monthly
7	Product quality is measured and reported using appropriate statistical methods.	Not applicable	Not applicable	Not applicable
8	Each day, cross-functional teams analyze the process and product data and look for improvement opportunities.	Participate in process improvement teams, provide expert consulting when needed	Self, operations team	Monthly/as needed
9	Production schedules are easily maintained because surprise quality defects are nonexistent, and each operation is equipped to make changeovers rapidly.	Participate in SMED teams to reduce changeover times, develop engineered solutions for SMED-related equipment modifications	Self, operations team	As needed
10	Employees have the freedom from fear to try to make things better.	Seek new approaches to improve equipment efficiencies, offer suggestions to improve the system, assist in development of junior engineers	Self	As needed

Table 13.11 Reliability Crew Leader's Action Plan

Vision Element		*How*	*Who*	*When*
1	Workplace is clean, well lit, properly ventilated, and free of safety hazards.	Make daily walk throughs of assigned areas and job sites, perform frequent informal STOP observations and formal monthly STOP observations	Self	Daily/ monthly
2	Machines are in tip-top mechanical condition and capable of running without breakdown from one maintenance downtime to the next.	Facilitate/direct crew in performance of preventive (PM), predictive (PdM), and operator-assisted (TPM) maintenance tasks, ensure work orders are planned and executed per established operations team priorities	Self, reliability crew, reliability planner	Daily
3	Process is monitored and controlled with modern instrumentation.	Monitor work order system to ensure maintenance and calibration of process instrumentation per plan	Self, reliability crew, reliability planner	Daily
4	Only high-quality raw materials are used.	Not applicable	Not applicable	Not applicable
5	Workforce is trained to be "highly skilled professionals."	Coordinate training of crew members to develop expertise on assigned processes and associated equipment, request specialized training from equipment manufacturers as needed	Self, reliability engineer, reliability crew, equipment manufacturers	Ongoing

6	Operators monitor product characteristics using standardized methods and adjust settings using SPC techniques.	Respond to requests for assistance in resolution of variability issues	Self, reliability crew, reliability engineer, process engineers	As needed
7	Product quality is measured and reported using appropriate statistical methods.	Not applicable	Not applicable	Not applicable
8	Each day, cross-functional teams analyze the process and product data and look for improvement opportunities.	Attend daily operations meeting, make crew members available to participate in process improvement teams, provide expert consulting when needed	Self, reliability crew, operations team	As needed
9	Production schedules are easily maintained because surprise quality defects are nonexistent, and each operation is equipped to make changeovers rapidly.	Make crew members available to participate in SMED teams to reduce changeover times, purchase or manufacture special jigs and fixtures, install on process equipment as needed	Self, reliability crew, operations team	As needed
10	Employees have the freedom from fear to try to make things better.	Employee advocacy, assistance, facilitate employee suggestions	Self, crew members	Daily

Table 13.12 Reliability Planner's Action Plan

Vision Element		*How*	*Who*	*When*
1	Workplace is clean, well lit, properly ventilated, and free of safety hazards.	Ensure shop is stocked with all required safety equipment and that annual testing and certification of equipment are performed by licensed contractors	Self	Daily
2	Machines are in tip-top mechanical condition and capable of running without breakdown from one maintenance downtime to the next.	Develop expertise on CMMS, develop equipment bills of materials, develop and implement PM schedule, develop library of preplanned jobs, incorporate technician feedback into job plans, assess returned work orders for completeness	Self, reliability crew	Ongoing
3	Process is monitored and controlled with modern instrumentation.	Use CMMS to schedule maintenance and calibration of process instrumentation per plan	Self, reliability crew	Daily
4	Only high-quality raw materials are used.	Not applicable	Not applicable	Not applicable
5	Workforce is trained to be "highly skilled professionals."	Attend CMMS training, provide CMMS training to technicians, incorporate technician training time into work order planning	Self, reliability crew leader, CMMS vendor	As needed

6	Operators monitor product characteristics using standardized methods and adjust settings using SPC techniques.	Schedule technicians to respond to requests for assistance in resolution of variability issues as directed	Self, reliability crew leader, technicians	As needed
7	Product quality is measured and reported using appropriate statistical methods.	Not applicable	Not applicable	Not applicable
8	Each day, cross-functional teams analyze the process and product data and look for improvement opportunities.	Attend daily operations meeting, schedule crew members to participate in process improvement teams as directed	Self, reliability crew, operations team	Daily
9	Production schedules are easily maintained because surprise quality defects are nonexistent, and each operation is equipped to make changeovers rapidly.	Schedule crew members to participate in SMED teams as directed, purchase or initiate work orders to manufacture special jigs and fixtures, initiate work orders to install on process equipment as requested	Self, reliability crew, operations team	As needed
10	Employees have the freedom from fear to try to make things better.	Provide assistance to reliability team and others, look for ways to streamline system	Self	Daily

Table 13.13 Reliability Crew Member's Action Plan

Vision Element		*How*	*Who*	*When*
1	Workplace is clean, well lit, properly ventilated, and free of safety hazards.	Keep shop and job site areas clean and free of hazards, employ safety procedures like lock and tag and vessel entry, perform work using appropriate tools and prescribed safety equipment	Self	Daily
2	Machines are in tip-top mechanical condition and capable of running without breakdown from one maintenance downtime to the next.	Complete assigned work orders using good workmanship, add work order completion information needed to improve job plan, make daily walk-throughs of all assigned process equipment and informal assessment of mechanical condition, perform PdM tasks per plan via work order, respond promptly to all breakdowns and requests for assistance, provide TPM operator training as assigned	Self	Daily
3	Process is monitored and controlled with modern instrumentation.	Perform maintenance and calibration of process instrumentation per assigned work orders	Self, planner	Daily
4	Only high-quality raw materials are used.	Not applicable	Not applicable	Not applicable
5	Workforce is trained to be "highly skilled professionals."	Develop expertise on assigned processes and associated equipment through process control manuals and manufacturer's literature, request specialized training from equipment manufacturers as needed	Self, reliability engineer, reliability crew leader, equipment manufacturers	Ongoing

6	Operators monitor product characteristics using standardized methods and adjust settings using SPC techniques.	Respond to requests for assistance in resolution of variability issues as directed by crew leader	Self, reliability crew leader, reliability engineer, process engineers, operators	As needed
7	Product quality is measured and reported using appropriate statistical methods.	Not applicable	Not applicable	Not applicable
8	Each day, cross-functional teams analyze the process and product data and look for improvement opportunities.	Participate in process improvement teams as directed by crew leader, provide expert consulting as needed	Self, operations team	As needed
9	Production schedules are easily maintained because surprise quality defects are nonexistent, and each operation is equipped to make changeovers rapidly.	Participate in SMED teams to reduce changeover times, assist in development of special jigs and fixtures, install on process equipment as directed	Self, operations team	As needed
10	Employees have the freedom from fear to try to make things better.	Offer suggestions to improve system, assist junior employees in their deployment	Self	Daily

up at your house to make a repair. This value is increased by actively working to improve the system through suggestions and participation in improvement teams.

Summary

The intent of this chapter is to put together a matrix of actions such that, if carried out, they would enable an organization to collectively pursue each element of the Model Vision day in and day out. I tried to be as brief and succinct as possible and only touch on the high points. Once you immerse yourself in the process, many other details will become known, and you can add them to the action plans. That way, they will not be forgotten, and over time the action plans will be perfected for your particular organization and style of management.

Final Comments

In closing, I would like to underscore my firm belief that making other people successful by removing barriers is rewarding from both a personal and a business standpoint. But, you cannot figure out what those barriers are by sitting in your office. When you spend some time out on the plant floor, you will soon find that people really do want to take pride in their work and when given the chance will nearly always exceed expectations. So, when the word gets out that you enjoy helping people, you will have plenty of opportunities, and that is when the management job finally becomes fun.

Chapter 14

A Case Study

Some years ago, I had the opportunity to apply Deming's principles with a free hand. One of the manufacturing departments at the plant where I was working had undergone a multimillion-dollar "upgrade" but would not perform. The upgrade was intended to increase capacity, but the process ran so badly after the project was completed that output was considerably less than before. I was given the assignment to get the department running and due to the urgency was told to do whatever was necessary to make it happen.

While I was initially flattered that management would place that kind of trust in my abilities, I was also terrified because that department was in terrible shape. Anyone casually walking through it could see that it was not actually "running." None of its five production lines ran continuously. It was more a series of starts and stops. Quality was terrible, and there were frequent injuries as people rushed around doing their best to try to keep the lines running.

I felt sure this operation could be turned around, but the situation was dire. The department had an operating budget in today's dollars of about $10 million per month. Compared to standards, it was losing over a million dollars per month. It was daunting, to say the least. Fortunately, I had just been through the Deming videotape series, and it was fresh in my mind, so I at least had a road map. But, did I have the time to put it in place?

The department began each day with a morning meeting. Initially, ours were all about the same—a rehash of what had broken down the day before. I repeatedly asked why we had all the breakdowns and eventually got to the truth. To "save money," maintenance had been neglected for years, which explained why everyone in the department was so demoralized.

They had been fighting a losing battle for quite a while. Also, it had become ingrained that they could not spend any real money, so they had been patching things together instead of replacing worn-out components. There is an old rule of thumb in maintenance that you should be spending about the same amount of money for parts as you do for labor. From the monthly financial reports, they clearly had not been buying enough parts.

I set a new ground rule. I said that with so many things that needed to be fixed, we only have time to fix something once, so whatever we decide to do has to be a permanent solution—regardless of cost. After the morning meeting, I would go out to the department and just stand around and watch. After a couple of days of doing this, one of the lead operators approached me and said, "You're the new boss, aren't you?" Yes, I replied and introduced myself. He said, "I have something I'd like to show you." I said, "Sure, let's take a look." He walked over to his machine and showed me a section of machine-indexed flight conveyor that kept falling apart and asked if something could be done about it. I said, "Wait here. I'll be right back." I went and found one of the engineers and the maintenance supervisor and asked them about the problem. They said they knew about it but had not been able to find a locknut that did not eventually back off. I said, "What if we drill the bolts and install cotter pins?" They allowed that it would take some time but should work. I took them over to speak to the machine operator, and they told him what they planned to do. By the end of the day, they had built a replacement assembly in the shop and installed it on the machine. I went back to the operator the next day and said, "Well, did they get your problem solved?" He said, "They sure did! Now I have something else to show you." And so began our new adventure.

Each day, I would go out and direct the team just like a coach on the sidelines. Our list of things to fix quickly piled up, and we began to set priorities. Within two months, I had authorized purchase of what would be a million dollars today in new parts. The team responded like the proverbial kid in a candy shop. Now things were getting fixed, but the amount of money I was spending blew our maintenance budget through the roof. It did not go unnoticed. However, I kept the heat I was getting from the front office from the team. They needed to be free from fear even if I was not.

Eventually, things began to improve, so I told the team that they did not need me to direct traffic any more. They knew what to do and should spend their time on the floor with the operators looking for problems. I would be there for them as a consultant any time they wanted to stop by my office. Those one-on-one consultations were a joy. Someone would stop by and tell me about a problem they were working on, what they had been thinking about for a solution, and what it would cost. Likely as not, I had nothing to add but occasionally had a suggestion or two, and off they went. We were finally starting to have some fun.

While the process was running considerably better, I was not convinced the main machines, which had been neglected for so long, were in optimum condition. It still took too much fiddling to keep them going. The operating window was much too small, and unless the planets were exactly aligned, they would get "out of time," which necessitated more fiddling and downtime. I called the factory and asked if they could send one of their best technicians to our plant and restore one of our machines literally to like-new condition in place. They said they could. The factory technician arrived, and we took one of the lines down for a month for a rebuild. This cost a fortune, but it was worth it. When we started the machine again, it ran as smooth as glass, and you did not have to beg it to run because it was in perfect alignment and all of the mechanism tolerances had been restored to factory-new conditions. I said to the team, "This proves it can be done. Can you duplicate this on the other lines?" "No problem!" they replied. And they did.

With nearly all of the recurring failures behind us, it was time to move to the next level. One of the things we got from the department upgrade project was programmable logic controllers (PLCs) in place of relay logic. This allowed us to access their memories and actually see which components on the lines were causing the most delays. I took the data, made Pareto charts, and had them posted on the department bulletin boards to let the employees know what we were working on. We took the top three delay items, put them on run charts, and went to work to try to reduce them. Now, we had to dig because we had fixed nearly all the obvious things, and the next levels were tougher. So, we began to spend time in meetings on fishbone diagrams or something similar to try to discover root causes of the things we could not see. By now, a year had passed. We had spent several million dollars on maintenance and upgrades, but the department was running about $6 million a year better in terms of productivity.

As our confidence began to build, it seemed like it might be a good time to show the Deming video series to the team. So, I began scheduling weekly sessions for us to watch them and discuss what we learned. This was a group of 20-somethings, and yet they loved the lectures as much as I did. They clearly looked forward to these sessions, and if we had to miss a week due to a schedule conflict, they let me know their disappointment.

Somewhere along the line, without being fully aware of it, we began shifting our attention from machine problems to dealing with quality defects. This was probably driven by the fact that, even though our productivity levels were coming into line, our defect rate was still high, and we were dead last compared to our three sister plants. So now in addition to the Pareto charts and productivity charts, I posted the percentage defective by category and comparisons to the other plants. In response, the team began focusing on quality and digging out the root causes of the defects—literally getting the red beads out of the system. And along the way, a funny thing happened. As we improved our quality from last place to first place, our productivity and cost of manufacturing improved by what would be another $6 million per year in today's dollars, making us the best in the company. This did not happen overnight. It took another year, but we did it.

All of that happened at a union plant, which clearly made no difference in terms of human response. When we removed the barriers that kept people from taking pride in their work—they took pride in their work. You could see it in their faces, and a number of production records were set without management putting up any signs or banners about increasing productivity.

When I was transferred from that facility to another location, the quality manager presented me a framed copy of our percentage defective trend charts at my farewell party. The charts showed continuous improving trends over a two-year period and approached near zero at the end. That was my trophy—and one I treasured—because it represented recognition of my work by my trusted friend and colleague.

After many years, I look back at those days with a lot of satisfaction. The assignment was difficult and at times somewhat terrifying. Without Dr. Deming's road map, I am not sure I would have had the confidence to plow ahead. One thing is certain: The process changed a number of lives for the better.

Appendix 1: The Red Bead Experiment*

Introduction

This experiment can be conducted with a Lightning Calculator Sampling Bowl or Sampling Box. While Dr. Deming used 80% white beads and 20% red, you will find this experiment done with 80% white and 20% colored. The colored beads add another dimension to the demonstration. This will be become self-evident as the script is read.

This is a clever demonstration of the futility of most management systems for improving quality. Dr. Deming often refers to it as a stupid experiment that you'll never forget. The experiment is described in a form similar to Dr. Deming's presentation in his seminars. As will be described at the end, it can also be adopted for very small groups and even a one on one presentation.

The experiment starts with a sampling device that has 80% white and 20% colored beads, normally red beads, hence the name "red bead experiment." The fact that the some sampling devices have various colored beads instead of all red is of no consequence, the experiment works just as well. It is necessary however to have 20% colored beads to go along with the text of this demonstration. If different colored beads are used instead of all red, then the colors can represent different kinds of defects. (This will become clearer as the demonstration is described.)

The objective of the bead factory in the demonstration is to make white beads. The customer will not accept anything but white beads, all colored beads are defective. The colored beads themselves represent defects in an organization's business processes. They represent a faulty machine or tool, a

* Courtesy of Lighting Calculator (http://www.qualitytng.com).

bad engineering design, a defective part, a procedural flaw, an unreasonable change request, ... all the things that can and do go wrong with a process. Supervisors and management control the number of red beads in the processes that are given to the workers. Now, let's proceed assuming that management has developed and purchased the white bead process for the workers of this experiment.

The Experiment

The instructor should take on the role of the department foreman or department supervisor. First of all he selects his work team. Realizing that one of the objectives of this demonstration is to point out management prejudices, the instructor can use whatever slogans or phrases he believes fit his particular audience. It many start as below.

Foreman: Okay I need eight bodies. (Pointing to people he wants to select) Can you count? Okay you're hired. Can you push buttons? Okay you're hired. You don't have to think, you just have to do what I tell you. You'll be on an apprenticeship for a while, and if you work out we'll hire you. We believe in high quality. We need good people.

So the process goes until all nine people are selected. The roles they will perform are as follows:

Foreman (the instructor)
Worker 1—Bob (assuming fictitious names for this description)
Worker 2—Dorothy
Worker 3—Henry
Worker 4—Calvin
Worker 5—Carol
Worker 6—Judy
Inspector 1—Ron
Inspector 2—Marty
Chief Inspector—Darwin

What will happen is that each worker will take turns drawing a sample of 50 from the sampling device. If you are using a sampling bowl then use the 50 hole paddle. If you are using a sampling box then designate the 50 hole pattern that you will use. This can be done verbally, mark on the face of the box with a water based transparency marker, or mask off the excess holes with masking tape, etc. Once a sample is drawn, then it will be checked by both inspectors (this high quality company has 200% inspection) who will independently write down the number of colored beads they count and show it to the chief inspector. The chief inspector will compare the counts, record the information on a data sheet and a graph, and then announce the number of colored beads drawn in the sample. The data sheet will show the names of the six workers and how many colored beads they each draw for the four days of the experiment. The data sheet will look like the sample data shown in Figure A1.1. The graph will have six plot points for each day for four days and look like the graph in Figure A2.2. These graphs can be set up on chart pads or overhead transparencies.

After the samples are counted, the beads are returned to the sampling device and another sample is drawn (shaking the box or mixing the bowl adds to the demonstration). Obviously the percent defective is a constant 20% but the actual percentage will vary with each sample due to sampling error and this is where the spoof begins. While there are some lessons in statistics than can be taught from this experiment, the real punch line is in the way the instructor conducts the demonstration and allows people to see themselves and the futility of management practices for improving quality. Each instructor has to know his audience and how far he can push his points without turning them off.

As the experiment is carried out over the four days the instructor uses the results of the samples to make his points. Realizing that the data will most likely vary between 1 and 19 (determined from the control chart calculations for UCL and LCL for a process that is 20% defective and sample sizes of 50). There are a variety of points that can be made. The exact order that the instructor will make the points will be dependent on the actual data developed in the experiment. However, after 24 samples you will have enough high and low readings along with increasing and decreasing trends to make all the points.

Clear Instructions

The first point of the demonstration has to do with giving clear instructions. Management often believes that if they make the objectives clear then the problems will go away. It could go something like the following:

Foreman: Bob you know that our job is to make white beads, the customer will not accept colored beads. You make white beads by first off making sure that the material is well mixed. If it is not mixed well you will have trouble making white beads. (Demonstrate how you want the box shaked or the beads mixed in the bowl.) You then turn the box over and hold it completely flat while you push this button on the end and white beads will fall into the holes. (If you are using a sampling bowl then the describe how the paddle is to be used to draw a sample.) Place the paddle into the bowl at the end, scoop deep into the beads and raise the paddle slowly at an angle of precisely 30 degrees and let the extra beads roll off the paddle. Now I've described the job for all you very clearly. I'm sure that all of you can now make white beads. Bob, would you please make our first batch of white beads?

Bob draws the first sample and shows it to the inspectors who record the count and they show it to the chief inspector. He then announces that we have 14 red beads. If colored beads are used then let red beads be red defects, green be green defects, yellow be yellow defects, etc. Still have the chief inspector count and record the total number of colored beads, but have him also report the different number for each color. The instructor can then use this to attach more blame to the worker, i.e., "Yellow defects, you know they are the worst kind and most costly to repair," and so on.

Foreman: Bob I've told you that the customer will only accept white beads, colored beads are not acceptable. Did you mix the material and hold the box completely flat like I told you? (Did you hold the sampling paddle at precisely 30 degrees?)

Bob: Yes I did.

Foreman: Well you must not have been paying attention. Dorothy can you please make us a batch of white beads? Remember all that I've told you. (Repeat the appropriate instructions again.)

Dorothy draws a sample and it comes out to be 12 colored beads.

Foreman: Well Dorothy that is better than Bob, but you must not have been paying attention either. Let me repeat the instructions once again. (Repeat the appropriate instructions.) Henry will you please run this process. Remember that the customer will only accept white beads.

Intimidation

Henry then draws a sample and it comes out to be 15.

Foreman: Henry I need to talk to you. Didn't you hear me when I gave all these people instructions on how to make white beads? What were you doing at the time, dreaming of some date you were going to have with Judy? I thought you said you wanted a job. We bring you here, give you clear instructions, show you how to make white beads and you still don't do it. What's the matter with you? Now I'm telling you people, you all better start paying attention or I'll have to fire all of you. Calvin it's your turn.

Praise and Comparison

Calvin draws a sample and it comes out at 8 colored beads.

Foreman: Now that's much better. Calvin you are catching on. Calvin got the same instructions as the rest of you and he is now beginning to master the process. He was almost twice as good as Henry. We are going in the right direction now. Henry, you especially need to watch Calvin and see how he did it. In fact, the rest of you should all watch Calvin and see how he does it. Carol, it is your turn.

Banners and Slogans

Carol draws a sample and it turns out to be 10.

Foreman: Carol, I told you to watch Calvin. He knows how to do it. Now all of you listen up. We have firm quality standards at this factory. Didn't you read the quality first banner over the door of the factory? See that poster on the wall over there, it says "Satisfied customers are happy customers and that means they will buy more." Quality is critical to our survival as a company and you know what that means for all our jobs. This company has to get the silver star quality award. It is crucial to our success in the market place. Judy please show us how it's done.

Judy draws a sample and it comes out to be 6. The foreman then walks over and talks to the chief inspector.

Foreman: Darwin, here is a prime example of quality improvement. You see once I pointed out to everyone that we are really serious about quality at this factory, people started to improve. Judy ran the best batch of white beads that we've seen so far. I think that we need to put a poster by everyone's machine instead of just a few on the wall that we have. You know if we buy some of those quality first buttons we all could wear one and give one to every employee that improves. I'm sure that we can drive the point home more about quality. While today wasn't the best I'm sure we'll do better tomorrow. (While walking over to Judy) Judy, excellent job. Keep up the good work.

Incentives

Tomorrow comes and the foreman asks Bob to run his batch of white beads. Bob runs his sample and it comes out at 10.

Foreman: Bob you've done better than you did yesterday but we are still going in the wrong direction. I've talked this matter over with our management and we are going to institute a quality bonus for everyone who runs good parts. If you guys and gals will all do better we will have a big pizza party and bonus for everyone.

After the bonus scheme is in place Dorothy draws 5, Henry draws 6, and Calvin draws 8 colored beads. The foreman then talks to the audience as if they were the management. He is explaining the value of the bonus scheme.

Foreman: Well after we instituted the bonus scheme that I recommended things got better immediately. While things have been going up and down we are still much better than we were yesterday. I think that our people are finally getting the message and all our efforts with the new quality banners and quality first buttons are paying off. We are definitely on the road to zero defects. I think that we should design another button with a big zero in the middle that is surrounded by gold stars that we can present to our best employees. In fact I would like for one of you to speak at our employee meeting and tell them about this exciting new zero defects button program.

Blame

Carol and Judy then each draw 11 colored beads. The foreman is very upset that things are going bad after he has told management that things were improving. He then goes over to the chief inspector and discusses the problem with him.

Foreman: Darwin, I'm really upset. These people don't care. Here you give them a good job and show that it is possible to improve and what do they do, ignore you and make life easy for themselves. I know that our incentive scheme was working, look at the results. I think Carol and Judy are spending too much time talking to one another and not paying attention to their job. I think that I'll warn both of them about their performance and tell them if I catch them talking again that I'll have to discipline them.

Performance Appraisals

This process continues again and again until all 24 samples have been taken. The data is then summarized for display and the supervisor then rates everyone's performance. The data is then placed on an overhead projector or summarized on a chart pad.

Foreman: Bob, you started off bad and then improved slightly. You have got to do better. Dorothy, you started off bad, then you improved, then you fell off the wagon again, and then improved again. You need to pay closer attention to your work and gain more consistency. Henry, you started off terrible and then you finally caught on. You have the same problem as Dorothy, you need more consistency. Calvin, you're the best employee in the department, but you still have room for improvement. You could be the first employee to earn the zero defects button that the manager talked about at the employee meeting. Carol, your performance needs to improve. Your overall rating was good. You can do better if you stop all that talking with Judy I warned you about. Judy, you need to pay attention to your work. You started off good and something must have distracted you. I think it was all the talking with Carol.

The foreman should now take on the role of explaining the results to management as he turns to the audience. The focus can now be on the graph and the noting of the various trends. The same praise and blame is offered for the trends. An evaluation of the trends for each day and an overall comparison of each of the workers may be appropriate at this point, as you try to present the results with the best possible explanation. The supervisor would then promise that they can do better and explain all the new quality programs that are in place to address the problems, i.e., posters, etc. He would make special note that he had warned Judy and Carol and gave them unfavorable performance evaluations.

Discussion of the Experiment

The instructor then steps out of his foreman role and now asks questions as an instructor in order to get the class to evaluate what has happened.

What was the average number of defects for the experiment (X-Bar)? Take the total number of colored beads (220) and divide by 24. (X-Bar = 220/24 = 9.2 for this experiment) Ask them what they expected the average to be and why. Facilitate a discussion that gets them to see that the average has got to be 20% defective since we know that there was a fixed 20% defective all the time in the sampling device. Explain that "p bar" is merely the average percent defective for a given sample size. Given that we had 20% defective, and samples of 50, 20% of 50 is 10. In our example the actual average was 9.2. This will vary slightly with each demonstration.

Next show that a simple p-Chart would have control limits of 1 and 17. This is determined by using the following formulas:

$$\text{P Bar} = 220/6 * 4 * 50 = .18$$

$$\text{UCL or LCL} = \text{X Bar} \pm 3\ \textit{Sq Root}\ [(\text{X-Bar})(1 - \text{P Bar})]$$

$$9.2 \pm 3\ \textit{Sq Root}\ [(9.2)(.82)] =$$

or

$$9.2 \pm 8.24 = .96 \text{ to } 17.44$$

What this means is that given a 20% defective process, and with sample sizes of 50, that the number of colored beads will vary over 99% of the time between 1 and 17 just due to random chance. In the demonstration we knew that it was random chance because we controlled the experiment by virtue of a constant number of beads. In actuality, we have what Dr. Deming calls a "stable process," or a system that is varying only because of random chance. This random chance is also referred to as only being affected by "chance or common causes." Note that a stable process may still turn out faulty items.

While there are a number of options for discussing the results of the experiment, the instructor can discuss the following points independently or use the Deming's 14 points slide in the appendix as a guideline.

Using the 14 point slide as a reference, add the following comments for each respective point:

Dr. Deming's Fourteen Points

1. **Create constancy of purpose for improvement of product and service.** The red bead experiment is a process. The only way to actually improve the process was to change the number of red, or colored beads, in the container. While the process the data varied showing good and bad trends, the foreman was deceived into thinking his actions were impacting the process when in fact only random variation was taking place. The focus of improvement has to be on changing business processes, not blaming people.
2. **Adopt the new philosophy.** Management controls the number of red beads in a process initially since they design the products, purchase the machines or facilities, and establish the operating procedures. Once a system is installed, management has to find a way to reduce the number of red beads if they want to improve the process. The problem is, that in most cases management does not know where the red beads are, but the workers do. It is only through a joint effort between management and workers that processes can be improved.
3. **Cease dependence on inspection to achieve quality**. While two inspectors counted the number of red beads, you can note how often the count may have differed. Point out the non-value added role of the inspectors in actually changing the process.
5. **Improve constantly and forever every process for planning, production, and service.** The number of red beads in this experiment causing the level of variation was totally controlled by management. The workers were totally helpless in affecting the number of

red beads that were drawn. In real life however there usually are some things that the workers can do to create improvement. But according to Dr. Deming common or chance causes are controlled by management (usually 85% of the problem), and local faults or special causes (15%) are controlled by workers. Hence the major responsibility for process improvement (85%) lies with management changing the systems.

6. **Institute training on the job.** Point out how a simple understanding of p-chart statistics would have helped the foreman recognize that the process was really a stable process and that the variation he was experiencing was really normal and expected. While the foreman gave clear instructions of what was expected to the workers on the first day, this alone did not create quality. Management clearly telling workers what they want does not mean that workers can achieve it.
8. **Drive out fear.** If management really wants to learn where the red beads are in a process so they can remove them, they must be willing to listen and stop blaming workers for poor results. Did the intimidation or comparisons of the workers have any real effect on the outcome of the red bead experiment? Management thought so at the time, but in reality the results were still caused by random variation.
10. **Eliminate slogan, exhortations, and targets for the workforce.** In the same way that intimidation and comparisons between workers did not impact the outcome of the red bead experiment, neither did the slogans, buttons, or incentive schemes. The understanding and focus has to be on the process and its capabilities.
12. **Remove barriers that rob people of pride of workmanship. Eliminate the annual rating or merit system.** How do you think the workers felt about their performance appraisals? Do you think that it motivated them to do better? Has anyone ever had a performance appraisal where the comments were similar to the same mumbo jumbo and clichés that were fed to the workers in this experiment?
14. **Put everyone in the company to work to accomplish the transformation.** If management does not actually solicit employee input on solving business processes they are passing up a golden opportunity for improvement. Everyone at all levels of the organization has to be focused on quality improvement.

In summary, this demonstration can be a lot of fun while still making a number of dramatic points. The use of participants keeps people involved and active. If you get involved and act out the roles, the class will do likewise. The more the instructor customizes the experiment to the class the better it will be received. Use phrases that are common to the work place. Finally, the power of the experiment is only limited to the creativity of the instructor.

Adaptations of the Experiment

Small Audiences—The previous description was designed for classroom use. There may be situations where it would be advantageous to perform the experiment on a smaller scale and more quickly. To abbreviate the experiment assume that the conversations take place between the foreman and one employee. Explain the requirements and then let the employee begin sampling. As data will of course immediately show red beads, the various management solutions can be stated to the employee. Cover the following issues between each sample like intimidation, unfair comparisons, praise, banners and slogans, providing incentives, and blame. Record the results of each sample and then give the person a performance appraisal at the end. Discuss how the employee felt as you were explaining to him all the management solutions for improvement.

Use of a Customer—Instead of having two inspectors and a chief inspector let one person review the bead sample who will be the customer. Prearrange with the customer to act like an irate customer when he gets different colored beads, etc.

Rework Department—If you are using a sampling bowl and would like to make a point about the cost and inconvenience of rework pick another worker to be a rework operator. Give them a bottle of white out correction fluid and tell them to paint some of the red beads white. Note that this can be messy and you need to also have a bottle of correction fluid solvent and a rag available for cleanup of the operator and the beads when you are finished. Letting them work on three to four beads will be sufficient. Make the point about how poor a job they have done in rework. Show the production workers all the problems they have caused. Let the inspectors or customer comment on the quality of the rework job.

Bead Box versus Sampling Bowl—While both devices can be used for demonstration purposes, each has its own advantages. The bead box is of course totally self-contained and is easily transported. Students can handle and play with the box as it is passed around. The Sampling Bowl on the other hand is very visual and works well with large audiences. It provides flexibility in determining the colors and percentages as well as sample sizes, etc. As previously mentioned, different colored beads can be used to represent different types of defects and problems.

The Fourteen Obligations of Top Management

1. Create constancy of purpose for improvement of product and services.
2. Adopt the new philosophy.
3. Cease dependence on inspection to achieve quality.
4. End the practice of awarding business on the basis of price tag alone. Instead, minimize total cost by working with a single supplier.
5. Improve constantly and forever every process for planning, production, and service.
6. Institute training on the job.
7. Adopt and institute leadership.
8. Drive out fear.
9. Break down barriers between staff areas.
10. Eliminate slogans, exhortations, and targets for the work force.
11. Eliminate numerical quotas for the work force and numerical goals for management.
12. Remove barriers that rob people of pride of workmanship. Eliminate the annual rating or merit system.
13. Institute a vigorous program of education and self-improvement for everyone.
14. Put everybody in the company to work to accomplish the transformation.

Dr. W. Edwards Deming

Appendix 2: Introduction to Statistical Process Control Techniques*

Contents

Preface

by Marilyn K. Hart, Ph.D. & Robert F. Hart, Ph.D.

Quality Control Today

In this era of strains on the resources and rising costs of manufacturing, it becomes increasingly apparent that decisions must be made on facts, not just opinions. Consequently, data must be gathered and analyzed. This is where statistical process control (SPC) comes in. For over 70 years, the manufacturing arena has benefited from the tools of SPC that have helped guide the decision-making process. In particular, the control chart has helped determine whether special-cause variation is present implying that action needs to be taken to either eliminate that cause if it has a detrimental effect on the process or to make it standard operating procedure if that cause has a beneficial effect on the process. If no special-cause variation is found to be present, SPC helps define the capability of the stable process to judge whether it is operating at an acceptable level.

The strength of SPC is its simplicity. And with the use of Statit on the computer to make the calculations and to plot the charts, the simplicity becomes complete.

New Demands on Systems Require Action

Accountability with hard data, not fuzzy opinions, is being demanded. Existing processes must be examined and new ones discovered. The good news is that improved quality inherently lowers costs as it provides a better product and/or service. Statistical Process Control provides accountability and is an essential ingredient in this quality effort.

Statistical Process Control is not an abstract theoretical exercise for mathematicians. It is a hands-on endeavor by people who care about their work and strive to improve themselves and their productivity every day. SPC charts are a tool to assist in the management of this endeavor. The decisions about what needs to be improved, the possible methods to improve it, and the steps to take after getting results from the charts are all made by humans and based on wisdom and experience. Everyone should be involved in this effort!

Socratic SPC—Overview Q & A

So What Is Statistical Process Control?

Statistical Process Control is an analytical decision making tool which allows you to see when a process is working correctly and when it is not. Variation is present in any process, deciding when the variation is natural and when it needs correction is the key to quality control.

Where Did This Idea Originate?

The foundation for Statistical Process Control was laid by Dr. Walter Shew[h] art working in the Bell Telephone Laboratories in the 1920s conducting research on methods to improve quality and lower costs. He developed the concept of *control* with regard to variation, and came up with Statistical Process Control Charts which provide a simple way to determine if the process is in control or not.

Dr. W. Edwards Deming built upon Shew[h]art's work and took the concepts to Japan following WWII. There, Japanese industry adopted the concepts whole-heartedly. The resulting high quality of Japanese products is world-renowned. Dr. Deming is famous throughout Japan as a "God of quality."

Today, SPC is used in manufacturing facilities around the world.

What Exactly Are Process Control Charts?

Control charts show the variation in a measurement during the time period that the process is observed.

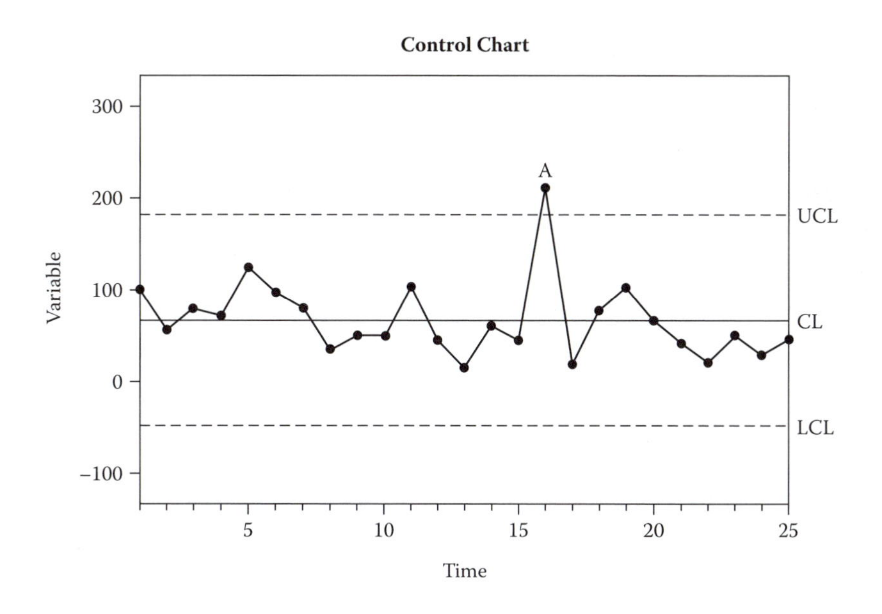

In contrast, bell-curve type charts, such as histograms or process capability charts, show a summary or snapshot of the results.

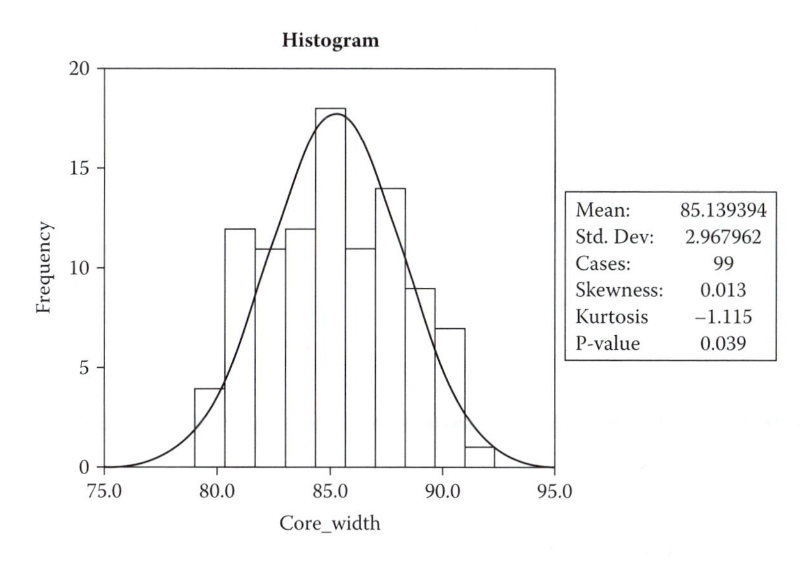

Process control charts are fairly simple-looking connected-point charts. The points are plotted on an x/y axis with the x-axis usually representing time. The plotted points are usually averages of subgroups or ranges of variation between subgroups, and they can also be individual measurements.

Some additional horizontal lines representing the average measurement and control limits are drawn across the chart. Notes about the data points and any limit violations can also be displayed on the chart.

What Is the Purpose of Control Charts?

Control charts are an essential tool of continuous quality control. Control charts monitor processes to show how the process is performing and how the process and capabilities are affected by changes to the process. This information is then used to make quality improvements.

Control charts are also used to determine the capability of the process. They can help identify *special* or *assignable causes* for factors that impede peak performance.

How Do They Work?

Control charts show if a process is *in control* or *out of control.* They show the variance of the output of a process over time, such as a measurement of width, length or temperature. Control charts compare this variance against upper and lower limits to see if it fits within the expected, specific, predictable and *normal* variation levels.

If so, the process is considered *in control* and the variance between measurements is considered normal random variation that is inherent in the process. If, however, the variance falls outside the limits, or has a run of non-natural points, the process is considered *out of control.*

What's This Relationship between Variation and Assignable Causes?

Variation is the key to statistical process control charts. The extent of variation in a process indicates whether a process is working as it should.

When the variation between the points is large enough for the process to be out of control, the variation is determined to be due to non-natural or *assignable* (special) causes.

So How Are These Normal-Predictable Variance Levels Determined?

One of the beauties of control charts is that the process itself determines the control limits. The process itself shows what can and cannot be expected. The control limits are automatically calculated from the data produced by the process. These calculations are done painlessly by Statit

QC software, no need to calculate them by hand. By definition control limits cannot be pre-assigned, they are a result of the process or the "voice of the process."

Control limits are NOT specifications, corporate goals, or the "voice of the customer." These two concepts must never be confused.

"Get real" might be a nice way of putting the concept of control limits vs. specification lines.

What about These Rules Violations *for Determining If a Series of Points within the Control Limits Is Unnatural?*

The work done by Shew[h]art and his colleagues gave them a base of empirical knowledge on which to base Rules Violations. For example, six points in a row steadily increasing or decreasing. These have been codified and are contained in the AT&T Statistical Quality Control Handbook. Statware uses these time-tested and industry proven standards to automatically check for rules violations. Full details of these rules are provided in AT&T's Statistical Quality Control Standards on page 235 and example charts on pages 94 to 96. They can also be found in the on-line help in Statit QC.

In Control? Out of Control? What's the Point?

If a process is in control, the outcomes of the process can be accurately predicted. In an out of control process, there is no way of predicting whether the results will meet the target. An out of control process is like driving a bus in which the brakes may or may not work and you have no way of knowing!

If a process is out of control, the next step is to look for the assignable causes for the process output, to look for the *out-of-controlness*. If this out-of-controlness is considered negative, such as multiple defects per part, the reasons for it are investigated and attempts are made to eliminate it. The process is continuously analyzed to see if the changes work to get the process back in control.

On the other hand, sometimes the out-of-control outcomes are positive, such as no defects per part. Then the assignable cause is sought and attempts are made to implement it at all times. If successful, the averages are lowered and a new phase of the process is begun. A new set of capabilities and control limits is then calculated for this phase.

What's This about Capabilities?

A control chart shows the capabilities of a process that is in control. The outcomes of the process can be accurately predicted, you know what to expect from the process.

Sometimes an organization's requirements, specifications or goals are beyond what the process is actually capable of producing. In this case, either the process must be changed to bring the specifications within the control limits, or the specifications must be changed to match the capabilities of the process. Other activities are a waste of time, effort and money.

Can Any Type of Process Data Be Judged Using Control Charts?

Processes that produce data that exhibits natural or common-cause variation will form a distribution that looks like a bell-curve. For these types of processes, control charts should provide useful information.

If the data is not normally distributed, does not form a bell-curve, the process is already out of control so it is not predictable. In this case we must look for ways to bring the process into control. For example, the data may be too broad, using measurements from different work shifts that have different process outcomes. Every process, by definition, should display some regularity. Organizing the data collection into rational subgroups, each of which could be in control, is the first step to using control charts.

OK, I'm Getting the Idea These Control Charts Are Key to Quality Improvement. What Specifically Do They Look Like, What Are Their Key Features and How Are They Created?

There are a handful of control charts which are commonly used. They vary slightly depending on their data, but all have the same general fundamentals.

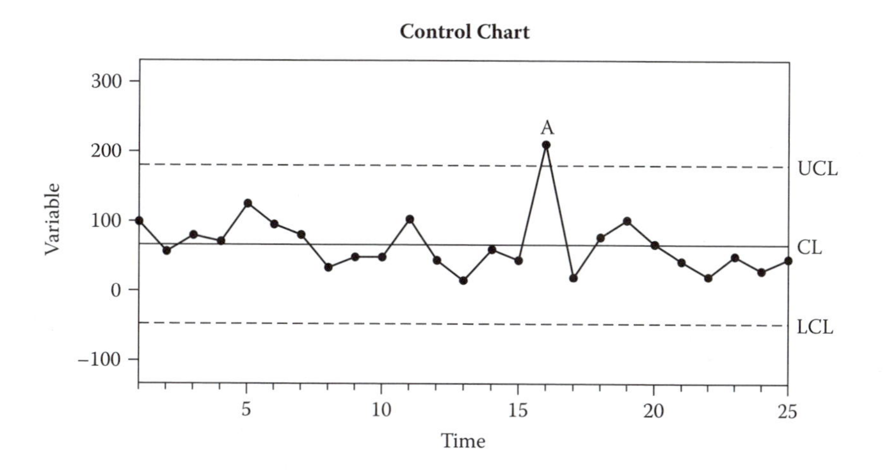

Control charts have four key features:

1) Data points are either averages of subgroup measurements or individual measurements plotted on the x/y axis and joined by a line. Time is always on the x-axis.
2) The Average or Center Line is the average or mean of the data points and is drawn across the middle section of the graph, usually as a heavy or solid line.
3) The Upper Control Limit (UCL) is drawn above the centerline and often annotated as "UCL." This is often called the "+3 sigma" line. This is explained in more detail in Control Limits on page 214.
4) The Lower Control Limit (LCL) is drawn below the centerline and often annotated as "LCL." This is called the "-3 sigma" line.

The x and y axes should be labeled and a title specified for the chart.

OK, Sounds Good. How Do I Go about Making These Control Charts?

In the past, creating control charts was a difficult process requiring statisticians and mathematicians to remember the formulas and actually calculate the various data points and control limits.

Today, using Statit QC software and computers, the complicated part of the task is done quickly and accurately. We can concentrate instead on the problems, solutions and increasing levels of quality, rather than poring over formulas. The time saved allows us to get immediate feedback on a process and take corrective actions when necessary. Our march toward Continuous Quality Improvement has never been less complicated.

Assuming measurements have been made and the data gathered, the next step is to open the data file, select the appropriate type of chart from the Statit menus, click the mouse and the chart is ready for inspection. Any out-of-control points have a letter over them. Click on the rule violation letter and a box is displayed describing the violation.

All Right! When Is the Best Time to Start?

NOW! The key to SPC is ***action!*** Rather than trying to discover the perfect methodology or data set, it is best to get started, to begin to truly effect quality immediately! Adjustments will be needed as you go along, this is a big part of the cycle of quality.

The cost of poor quality and non-conformance is great, in service industries it is estimated at 35%+ of gross revenue. Time, of course, is invaluable. Today, with the Statit, you can begin working immediately on quality improvement.

Steps Involved in Using Statistical Process Control

Proper Statistical Process Control starts with planning and data collection. Statistical analysis on the wrong or incorrect data is rubbish, the analysis must be appropriate for the data collected.

Be sure to PLAN, then constantly re-evaluate your situation to make sure the plan is correct.

The key to any process improvement program is the PDSA cycle described by Walter Shew[h]art.

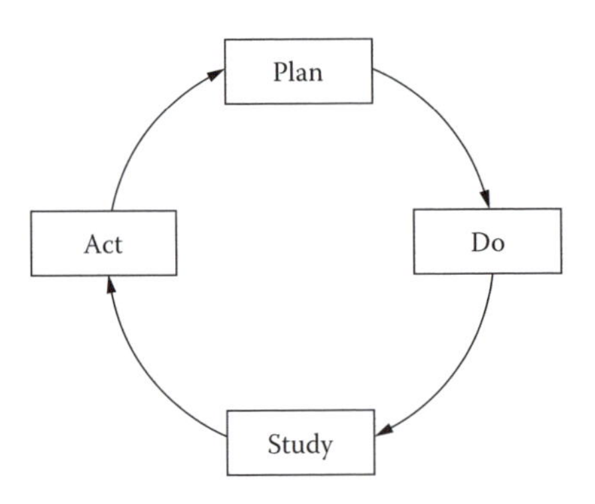

Plan

Identify the problem and the possible causes. The QC tools described in this manual can help organizations identify problems and possible causes, and to prioritize corrective actions.

Do

Make changes designed to correct or improve the situation.

Study

Study the effect of these changes on the situation. This is where control charts are used—they show the effects of changes on a process over time. Evaluate the results and then replicate the change or abandon it and try something different.

Act

If the result is successful, standardize the changes and then work on further improvements or the next prioritized problem. If the outcome is not yet successful, look for other ways to change the process or identify different causes for the problem.

Control charting is one of a number of steps involved in Statistical Process Control. The steps include discovery, analysis, prioritization, clarification, and then charting. Before using Statit QC software, appropriate data must be collected for analysis. Then, you need to begin again and do it over and over and over. Remember, quality is a CYCLE of continuous improvement.

Specific SPC Tools and Procedures

The preparatory phases of SPC involve several steps using a number of different tools. These tools are described below and most are available in Statit QC.

Eight quality tools are available to help organizations to better understand and improve their processes. The essential tools for the discovery process are:

Check Sheet
Cause-and-Effect Sheet
Flow Chart
Pareto Chart
Scatter Diagram
Probability Plot
Histogram
Control Charts
Brainstorming

Identification and Data Gathering

When you set out to improve quality, the first thing to do is identify the processes that need improvement. This can be done using a number of methods such as surveys, focus groups or simply asking clients about their experiences.

Once the problem areas are identified, a *brainstorming* session should occur with a variety of people who are involved with the processes. The target problems are decided upon and a list of possible causes is identified.

Prioritizing

After a number of possible problems are noted, the next step is to prioritize. The problems that are having the greatest effect are the highest priority items.

It has been "discovered" time and again that a great percentage of the trouble in nearly all processes is caused by a small percentage of the total factors involved. Service departments routinely find that 5% of the problems are taking over 80% of their time. Therefore, in order to maximize effectiveness, identify the key opportunities for improvement, those items that will provide the most benefit to your organization.

Pareto Charts

The Pareto chart can be used to display categories of problems graphically so they can be properly prioritized. The Pareto chart is named for a 19th century Italian economist who postulated that a small minority (20%) of the people owned a great proportion (80%) of the wealth in the land.

There are often many different aspects of a process or system that can be improved, such as the number of defective products, time allocation or cost savings. Each aspect usually contains many smaller problems, making it difficult to determine how to approach the issue. A Pareto chart or diagram indicates which problem to tackle first by showing the proportion of the total problem that each of the smaller problems comprise. This is based on the Pareto principle: 20% of the sources cause 80% of the problem.

A Statit QC Count Pareto chart is a vertical bar graph displaying rank in descending order of importance for the categories of problems, defects or opportunities. Generally, you gain more by working on the problem identified by the tallest bar than trying to deal with the smaller bars. However, you should ask yourself what item on the chart has the greatest impact on the goals of your business, because sometimes the most frequent problem as shown by the Pareto chart is not always the most important. SPC is a tool to be used by people with experience and common sense as their guide.

This is a Pareto chart of complaints from a customer satisfaction survey.

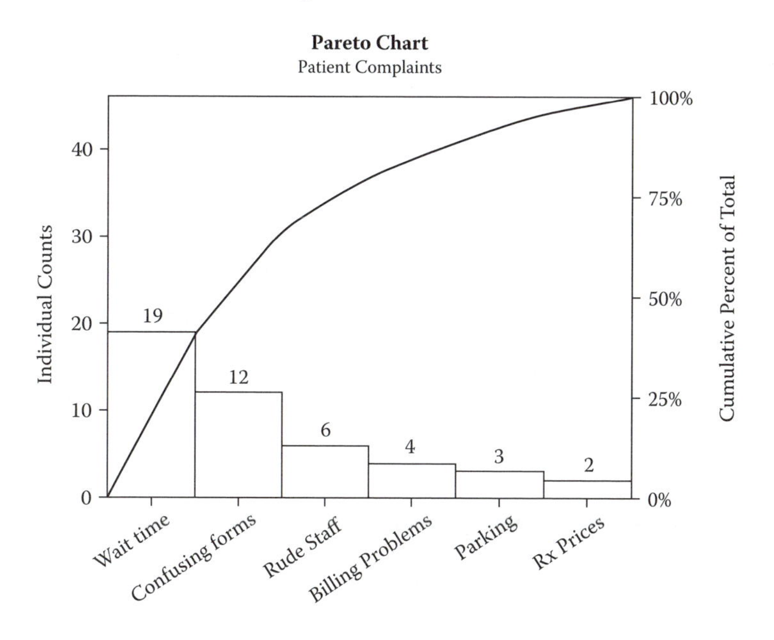

To create a Pareto chart:

- In Statit Express QC, click the **Pareto chart button** on the menu bar.
- In Custom QC, use the menu selection:
  QC → Attribute Charts → Pareto Chart

The dialogues lead you through the creation of the chart.

If you have any difficulties, open the Help file using the **Help** button on the Pareto Chart dialog, or select **Help → Contents** and find "Pareto Chart" in the Index.

Analysis of Selected Problem

Once a major problem has been selected, it needs to be analyzed for possible causes. Cause-and-effect diagrams, scatter plots and flow charts can be used in this part of the process.

Cause-and-Effect or Fishbone Diagram

One analysis tool is the Cause-and-Effect or Fishbone diagram. These are also called Ishikawa diagrams because Kaoru Ishikawa developed them in 1943. They are called fishbone diagrams since they resemble one with the long spine and various connecting branches.

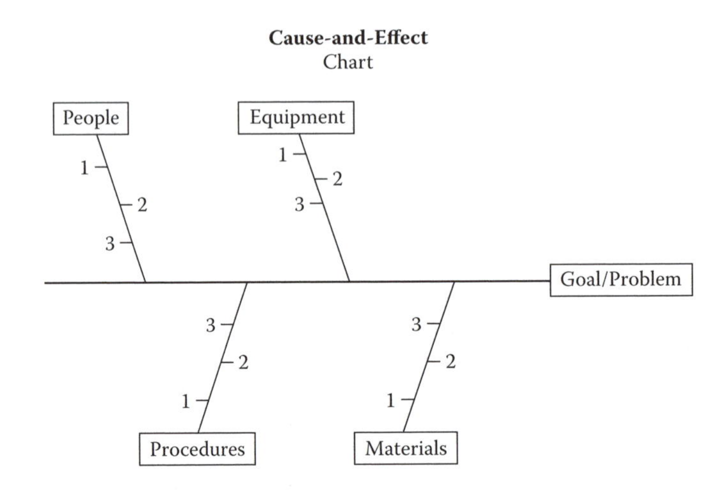

The fishbone chart organizes and displays the relationships between different causes for the effect that is being examined. This chart helps organize the brainstorming process. The major categories of causes are put on major branches connecting to the backbone, and various sub-causes are attached to the branches. A tree-like structure results, showing the many facets of the problem.

The method for using this chart is to put the problem to be solved at the head, then fill in the major branches. People, procedures, equipment and materials are commonly identified causes.

This is another tool that can be used in focused brainstorming sessions to determine possible reasons for the target problem. The brainstorming team should be diverse and have experience in the problem area. A lot of good information can be discovered and displayed using this tool.

To create a Fishbone diagram:

- In Statit Custom QC, use the menu selection:
  QC → Fishbone Diagram

The dialogs lead you through the creation of the chart.

If you have any difficulties, open the Help file using the **Help** button on the Fishbone Diagram dialog, or select **Help → Contents** and find "Fishbone Diagram" in the Index.

Flowcharting

After a process has been identified for improvement and given high priority, it should then be broken down into specific steps and put on paper in a flowchart. This procedure alone can uncover some of the reasons a process is not working correctly. Other problems and hidden traps are often uncovered when working through this process.

Flowcharting also breaks the process down into its many sub-processes. Analyzing each of these separately minimizes the number of factors that contribute to the variation in the process.

After creating the flowchart, you may want to take another look at the fishbone diagram and see if any other factors have been uncovered. If so, you may need to do another Pareto diagram as well. Quality Control is a continual process, in which factors and causes are constantly reviewed and changes made as required.

Flowcharts use a set of standard symbols to represent different actions:

Circle/Oval
  Beginning or end

Square
  A process, something being done

Diamond
Yes/No decision

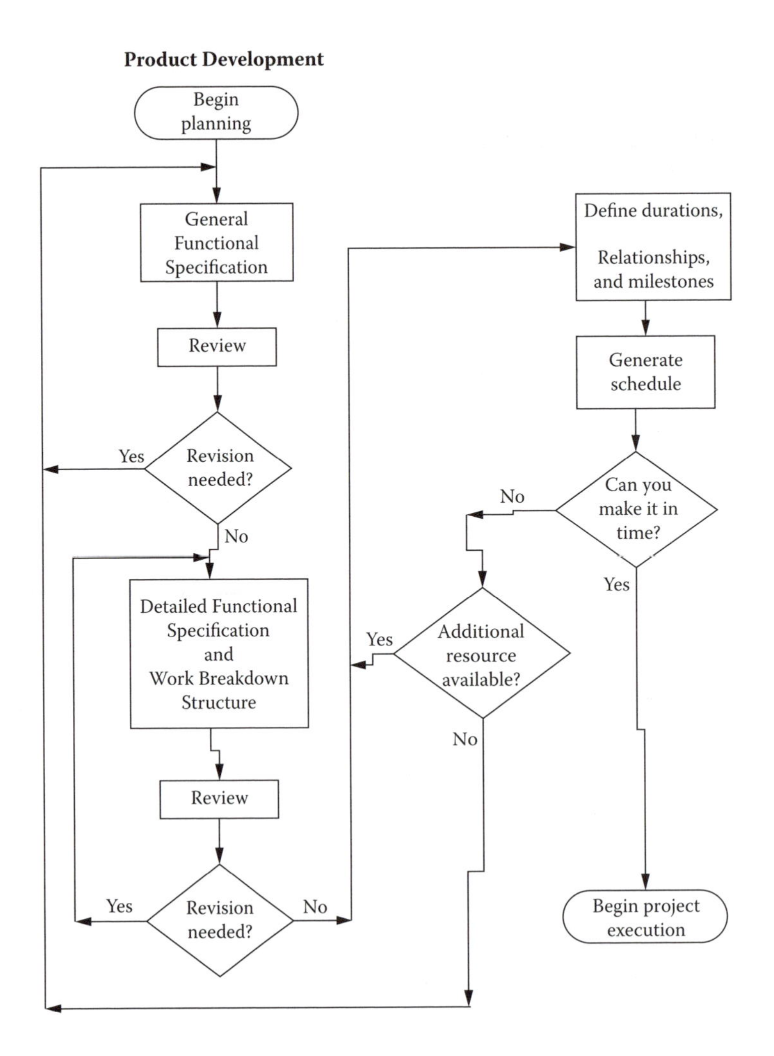

Scatter Plots

The Scatter plot is another problem analysis tool. Scatter plots are also called correlation charts.

A Scatter plot is used to uncover possible cause-and-effect relationships. It is constructed by plotting two variables against one another on a pair of axes. A Scatter plot cannot prove that one variable causes another, but it does show how a pair of variables is related and the strength of that relationship. Statistical tests quantify the degree of correlation between the variables.

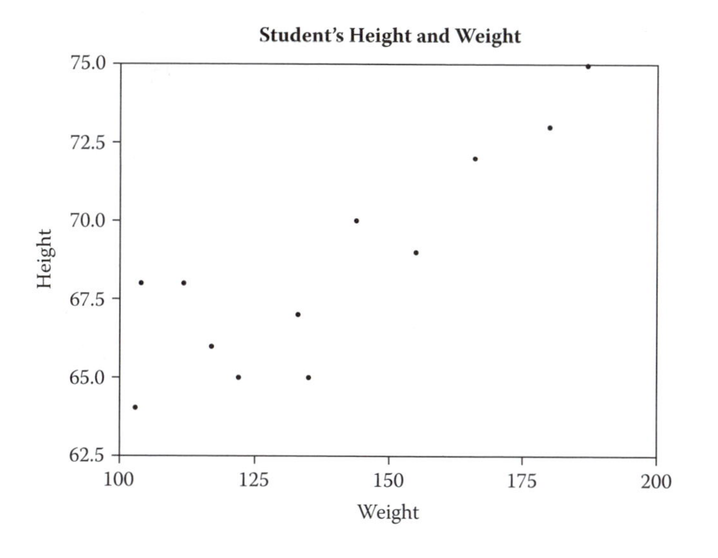

In this example, there appears to be a relationship between height and weight. As the student gets taller, generally speaking they get heavier.

To create a Scatter plot:

- In Statit Express QC, use the menu selection:
  Graphs → Scatter
- In Statit Custom QC, use the menu selection:
  Graphics → Relationship Plots → Scatterplot Matrix.

The dialogs lead you through the creation of the chart.

If you have any difficulties, open the Help file using the **Help** button on the Scatter Plot dialog, or select **Help → Contents** and find "Scatter Plot" in the Index.

Data Gathering and Initial Charting

This is the time to begin gathering data related to the problem. The following tools will help with this task.

Check Sheets

Check sheets are simply charts for gathering data. When check sheets are designed clearly and cleanly, they assist in gathering accurate and pertinent data, and allow the data to be easily read and used. The design should make use of input from those who will actually be using the check sheets. This input can help make sure accurate data is collected and invites positive involvement from those who will be recording the data.

Check sheets can be kept electronically, simplifying the eventual input of the data into Statit QC. Statit QC can use data from all major spreadsheets, including Excel and Lotus 123, all major database programs and some other SPC software programs. Since most people have a spreadsheet program on their desktop PC, it might be easiest to design a check sheet in a spreadsheet format.

Check sheets should be easy to understand. The requirements for getting the data into an electronic format from paper should be clear and easy to implement.

Histograms

Now you can put the data from the check sheets into a histogram. A histogram is a snapshot of the variation of a product or the results of a process. It often forms the bell-shaped curve which is characteristic of a normal process.

The histogram helps you analyze what is going on in the process and helps show the capability of a process, whether the data is falling inside the bell-shaped curve and within specifications. See Process Capability Chart—cp on page 227 for more information.

A histogram displays a frequency distribution of the occurrence of the various measurements. The variable being measured is along the horizontal x-axis, and is grouped into a range of measurements. The frequency of occurrence of each measurement is charted along the vertical y-axis.

Histograms depict the central tendency or mean of the data, and its variation or spread. A histogram also shows the range of measurements, which defines the process capability. A histogram can show characteristics of the process being measured, such as:

- Do the results show a normal distribution, a bell curve? If not, why not?
- Does the range of the data indicate that the process is capable of producing what is required by the customer or the specifications?
- How much improvement is necessary to meet specifications? Is this level of improvement possible in the current process?

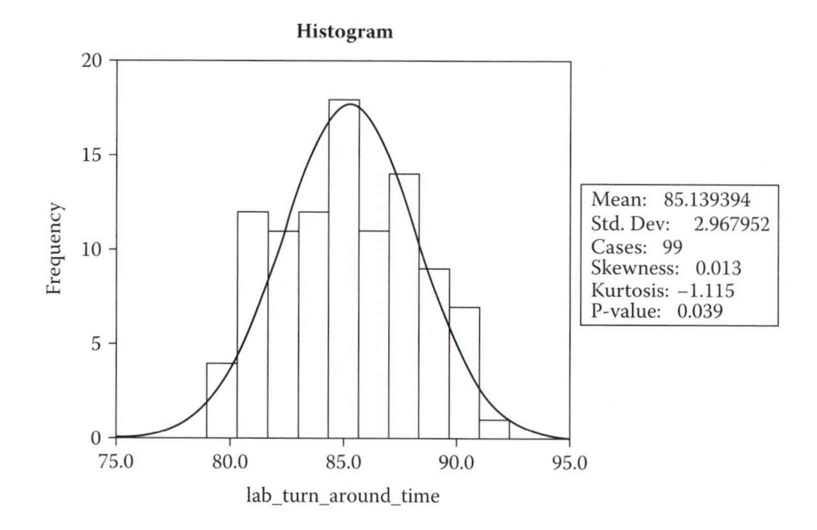

To create a Histogram:

- In Statit Express QC, use the menu selection:
  Graphs → Histogram
- In Statit Custom QC, use the menu selection:
  Graphics → Frequency Charts → Histogram …

The dialogs lead you through the creation of the chart.

If you have any difficulties, open the Help file using the **Help** button on the Histogram dialog, or select **Help → Contents** and find "Histogram" in the Index.

Probability Plot

In order to use Control Charts, the data needs to approximate a normal distribution, to generally form the familiar bell-shaped curve.

The probability plot is a graph of the cumulative relative frequencies of the data, plotted on a normal probability scale. If the data is normal it forms a line that is fairly straight. The purpose of this plot is to show whether the data approximates a normal distribution. This can be an important assumption in many statistical analyses.

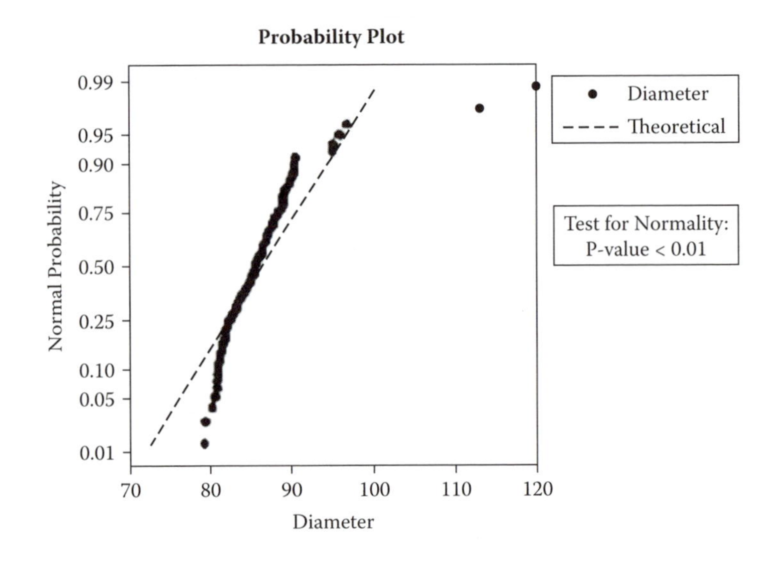

Although a probability plot is useful in analyzing data for normality, it is particularly useful for determining how capable a process is when the data is not normally distributed. That is, we are interested in finding the limits within which most of the data fall.

Since the probability plot shows the percent of the data that falls below a given value, we can sketch the curve that best fits the data. We can then read the value that corresponds to 0.001 (0.1%) of the data. This is generally considered the *lower natural limit*. The value corresponding to 0.999 (99.9%) is generally considered the *upper natural limit*.

Note: To be more consistent with the natural limits for a normal distribution, some people choose 0.00135 and 0.99865 for the natural limits.

To create a Probability plot:

- In Statit Express QC, use the menu selection:
  Graphs → Probability
- In Statit Custom QC, use the menu selection:
  Graphics → Distribution Plots → Probability ...

The dialogs lead you through the creation of the chart.

If you have any difficulties, open the Help file using the **Help** button on the Probability Plot dialog, or select **Help → Contents** and find "Probability Plot" in the Index.

The next section discusses control charts, which are considered the heart and soul of Statistical Process Control, and the main reason to use Statit QC.

The different types of control charts are explained in the next section, along with a discussion of control chart fundamentals.

Control Charts

Fundamental Concepts and Key Terms

Whether making mom's recipe for spaghetti sauce or admitting patients to an emergency room, the outcome of a process is never exactly the same every time. Fluctuation or variability is an inevitable component of all systems and is expected, arising naturally from the effects of miscellaneous chance events. However, variation outside a stable pattern may be an indication that the process is not acting in a consistent manner. Events which fall beyond expected variability or events forming a pattern that is not random, indicate that the process is out of control.

From a quality control perspective, an out-of-control service or production system is trouble! It is probably not be (sic) meeting customer specifications or achieving business goals, and there is no way of predicting if it will or can.

There are two general ways of detecting that a process is out of control. The first test for an out-of-control process asks, "Is any point falling above or below the control limits on its control chart?" This particular test is very easy to perform by viewing the control chart. The second form of rule violations is based upon patterns of points on the control chart and can be difficult to detect.

For example, 4 of 5 successive points in zone B or above, or 8 points on both sides of center with none in Zone C. Statit QC quickly and accurately performs these out-of-control tests, using predefined standard rule sets from AT&T's Statistical Quality Control Handbook, the definitive source for rule violation standards. See AT&T's Statistical Quality Control Standards on page 235. An even more detailed explanation is provided on pages 94 to 96.

Statistical Process Control charts graphically represent the variability in a process over time. When used to monitor the process, control charts can uncover inconsistencies and unnatural fluctuations. Consequently, SPC charts are used in many industries to improve quality and reduce costs.

Control charts typically display the limits that statistical variability can explain as normal. If your process is performing within these limits, it is said to be in control; if not, it is out of control.

It is important to remember what you can conclude about a system that is in control:

Control does not necessarily mean that a product or service is meeting your needs, it only means that the process is behaving consistently.

Rules Testing

How do you judge when a process is out of control? By plotting a control chart of the output of a process, it is possible to spot special or unnatural causes of variability and indications that the process is drifting. Drifting is defined by the mean or range of the variation changing as the process is running. The most common indication of change is a point falling outside of the control limits, but other tests for process instability are also valuable.

The rules used by Statit QC for control charts are based on rules described in AT&T's Statistical Quality Control Handbook. Different rules are appropriate for variable data and attribute data. Consequently, the default rules used by Statit QC vary with the type of chart being produced. These rules can be found in AT&T's Statistical Quality Control Standards on page 235. An even more detailed explanation is provided on pages 94 to 96, as well as in Statit QC's on-line help.

Zones in Control Charts

Many of the standard rules examine points based on *Zones*. The area between each control limit and the centerline is divided into thirds. The third closest to the centerline is referred to as Zone A, the next third is Zone B, and the third closest to the control limits is Zone C. Note that there are two of each of the Zones, one upper and one lower. See Control Limits on page 214 for more information.

Zone A is also referred to as the "3-sigma zone", Zone B is the "2-sigma zone", and Zone C is the "1-sigma zone". These sigma zone terms are appropriate only when 3-sigma is used for the control limits, as it is in Express QC.

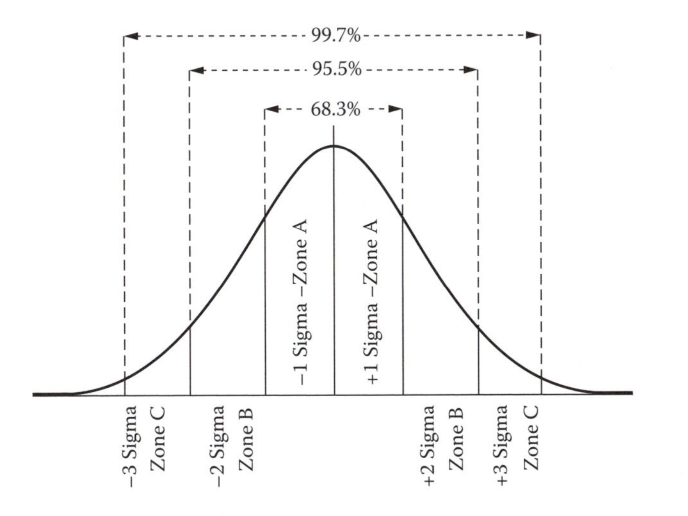

Sigma is the Greek letter for s and is used in this context to denote the spread of data.

Standard control limits are located 3 sigma away from the average or centerline of the chart. The centerline is also called the *control line.* These are called 3 sigma limits or 3 sigma zones. The distance from the centerline to the control limits can be divided into 3 equal parts of one sigma each.

Statistics tell us that in normal data dispersion, we can expect the following percentages of data to be included within the sigma:

1 sigma—68.3%
2 sigma—95.5%
3 sigma—99.7%

We can expect 99.7% of the process outcomes to be within the 3-sigma control limits.

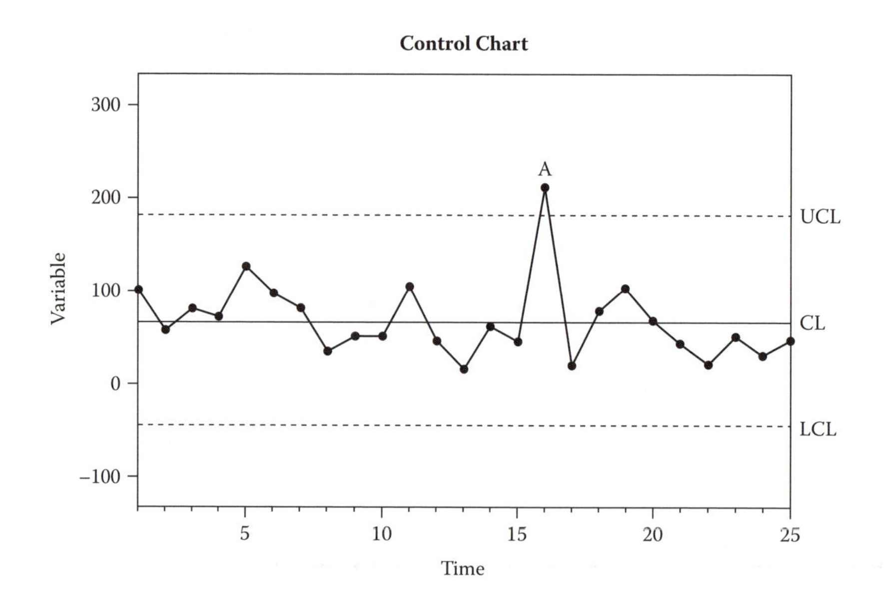

Control Limits

Control limits are calculated statistically from your data. They are referred to as the Lower Control Limit (LCL) and the Upper Control Limit (UCL) on a control chart. These are set at 3-sigma by default since this is the most commonly used limit.

Control limits define the zone where the observed data for a stable and consistent process occurs virtually all of the time (99.7%). Any fluctuations within these limits come from common causes inherent to the system, such as choice of equipment, scheduled maintenance or the precision of the operation that results from the design. These normal fluctuations are attributed to statistical variability.

An outcome beyond the control limits results from a *special cause.* Special causes are events external to the ordinary operation of a production or service. Special causes indicate that there have been one or more fundamental changes to the process and the process is out of control. Special causes need to be investigated and eliminated before a control chart can be used as a quality-monitoring tool.

The automatic control limits for Express QC have been set at 3-sigma limits. Warning limits are set at 2-sigma by default, and they can also be displayed.

Subgroups

An important factor in preparing for SPC charting is determining if you will measure every product of the process, such as measuring every part, or if you will use subgroups. Subgroups are a sample of data from the total possible data. Subgroups are used when it is impractical or too expensive to collect data on every single product or service in the process. Decisions to use subgroups or not [need] to be carefully thought out to ensure they accurately represent the data.

Subgroups need to be homogeneous within themselves so that special causes can be recognized, so problem areas stand out from the normal variation in the subgroup. For example, if you are in charge of analyzing processes in a number of facilities, a separate group should represent each facility, since each facility has different processes for doing the same tasks. Each facility subgroup should probably be broken down even further, for example, by work shifts.

Subgroups in Variable Control Charts

All data in a subgroup has something in common, such as a common time of collection, all data for a particular date, a single shift, or a time of day.

Subgroup data can have other factors in common, such as data associated with a operator, or data associated with a particular volume of liquid. In Express QC, this is referred to as a Grouped subgroup and there is a categorical variable that holds the grouping category, for example, Operator_ID or Volume.

Subgroups in Attribute Control Charts

A subgroup is the group of units that were inspected to obtain the number of defects or the number of rejects. The number of defects is displayed using c charts and u charts. The number of rejects, also called defective items, is displayed using p charts and np charts.

Phases

In Express QC, *phases* can be used in a variable control chart to display how a process behaves in response to a change in a particular characteristic of the system. For example, outside factors can cause disruptions in the normal process, such as a construction project in a plant that could cause additional defects.

Phase analysis can be helpful in identifying *special causes* of variation or trends. It is also useful when studying the impact of changes you make to your process, changes which you hope will be improvements.

Here is an example of a phase chart showing the changes that came about after new guidelines were established for shortening. You can see how the horizontal lines representing the upper control limit, centerline and lower control limit change with the new guidelines.

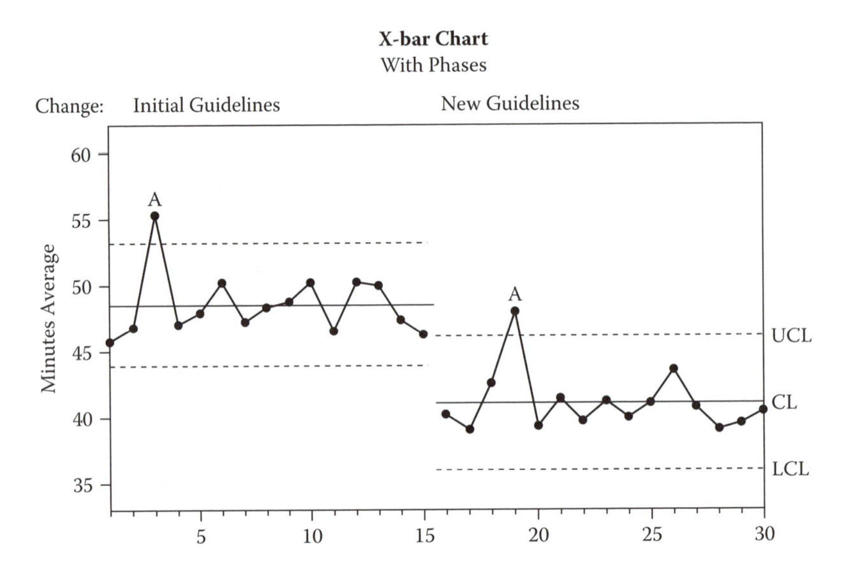

Sample Size

This is the number of cases making up the sample. The sample is a collection of observations used to analyze a system. In SPC applications, "Sample" is a synonym for "Subgroup."

Rejects—Nonconforming Items Data

Nonconforming items are rejects. A reject is tallied when an entire unit fails to meet acceptance standards, regardless of the number of defects in the unit. This includes defective products or unacceptable outcomes.

Defects—Nonconformities Data

Nonconformities are defects. A non-conformity is any characteristic, which should not be present but is, or a characteristic which needs to be present but is not. A defective item can have multiple non-conformities, for example, errors on insurance forms, incorrect medication, or service complaints.

Using Process Control Charts

OK, enough talk. Let's do some actual control charting. First you need to determine what data you have and select the correct chart for that data. Then make the chart and analyze it to see if the process is in control.

Data Definitions for Proper Chart Selection

Choosing the correct chart for a given a situation is the first step in every analysis. There are actually just a few charts to choose from, and determining the appropriate one requires following some fairly simple rules based on the underlying data. These rules are described in the flowchart below:

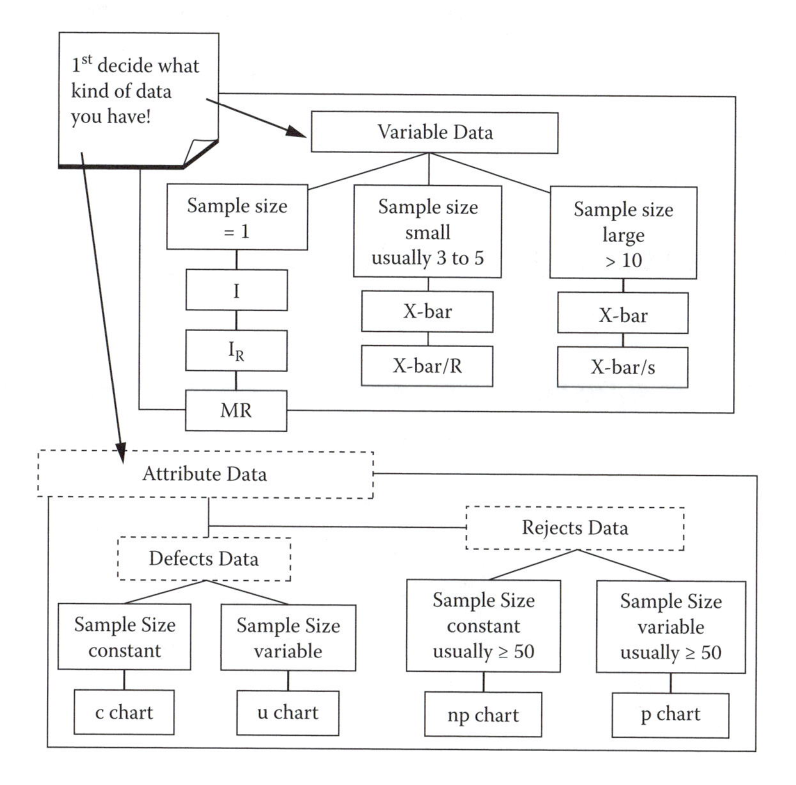

Control charts are divided into two groups:

Variable Data

Variable charts are based on variable data that can be measured on a continuous scale. For example, weight, volume, temperature or length of stay. These can be measured to as many decimal places as necessary.

Individual, average and range charts are used for variable data.

Attribute Data

Attribute charts are based on data that can be grouped and counted as present or not. Attribute charts are also called count charts and attribute data is also known as discrete data. Attribute data is measured only with whole numbers. Examples include:

Acceptable vs. non-acceptable
Forms completed with errors vs. without errors
Number of prescriptions with errors vs. without

When constructing attribute control charts, a subgroup is the group of units that were inspected to obtain the number of defects or the number of defective items.

Defect and reject charts are used for attribute data.

Types of Charts Available for the Data Gathered

Variable Data Charts—Individual, Average and Range Charts

Variable data requires the use of variable charts. Variable charts are easy to understand and use.

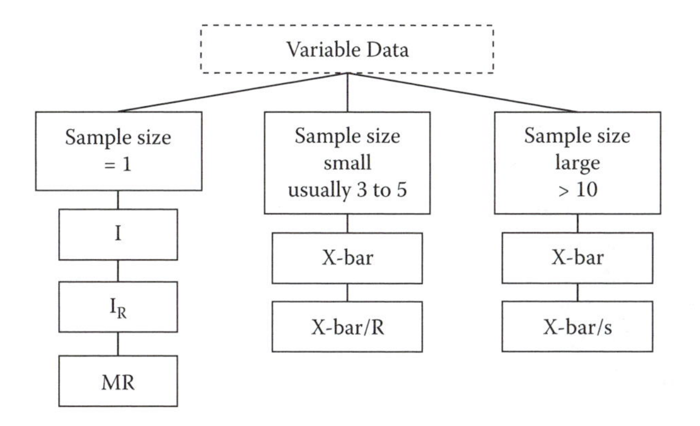

Individual Charts—I Chart

The **I chart** is also referred to as an individual, item, i, or X chart. The X refers to a variable X.

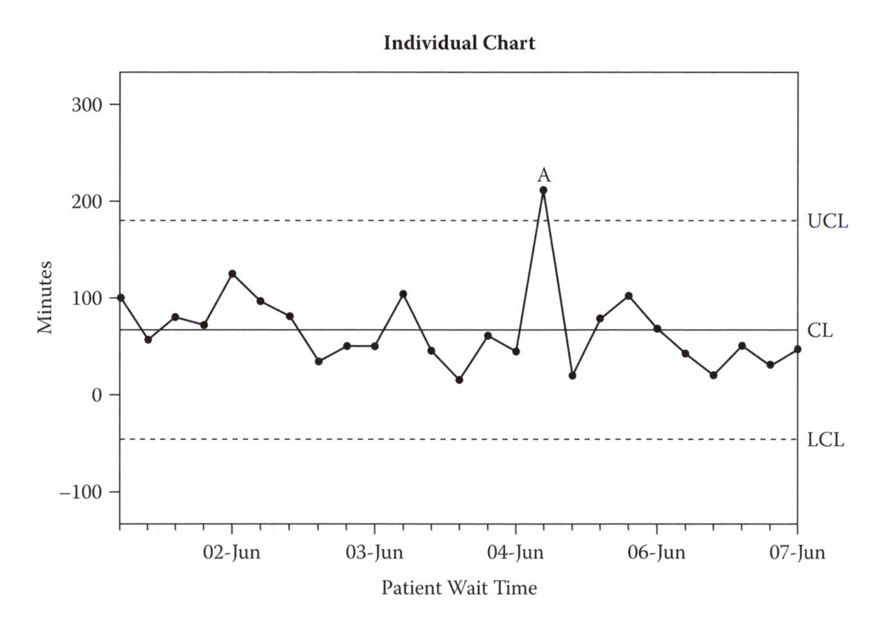

Individual charts plot the process results varying over time. Individual observations are plotted on the I chart, averages are not plotted on this type of chart. Individual charts are used to plot variable data collected chronologically from a process, such as a part's measurement over time.

These charts are especially useful for identifying shifts in the process average. When monitoring a system, it is expected that equal numbers of points will fall above and below the average that is represented by the centerline. Shifts or trends can indicate a change that needs to be investigated.

The individual control chart is reserved for situations in which only one measurement is performed each time the data is collected, where it is impractical or impossible to collect a sample of observations. When there are not enough data points to calculate valid control limits, an individual chart functions as a simple run chart.

To create an Individual chart:

- In Statit Express QC, click on the **"I"** button on the menu bar.
- In Statit Custom QC, use the menu selection:
  QC → Variable charts → Individual Chart

The dialogs lead you through the creation of the chart.

If you have any difficulties, open the Help file using the **Help** button on the Individual Chart dialog, or select **Help → Contents** and find "Individual Chart" in the Index.

Average Charts—X-bar Chart

Average charts are made by simply taking the averages of a number of subgroups and plotting the averages on the chart. The average chart is called the X-bar chart because in statistical notation, a bar or line over the variable (X) symbolizes the average of X.

"X-bar" is a shorthand way of saying "the average of X."

An X-bar chart is a variable control chart that displays the changes in the average output of a process. The chart reflects either changes over time or changes associated with a categorical data variable. The chart shows how consistent and predictable a process is at achieving the mean.

X-bar charts measure variation between subgroups. They are often paired with either Standard Deviation (S) or Range (R) charts, which measure variation within subgroups.

Definition: Variable Data Subgroups

All data in a subgroup has something in common, such as a common time of collection. For example, all data for a particular date, a single shift, or a time of day.

Subgroup data can have other factors in common, such as data associated with a particular operator, or data associated with a particular volume of liquid. In Express QC, this is referred to as a Grouped subgroup and there is a categorical variable that holds the grouping category. For example, Operator_ID or Volume.

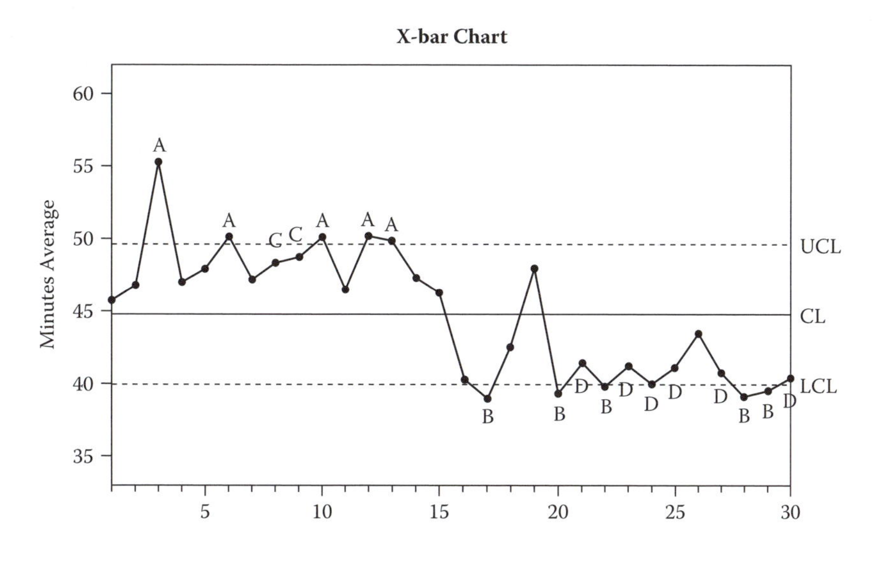

To create an X-bar chart:

- In Statit Express QC, click on the **"X-bar"** (a bar over the X) button on the menu bar.
- In Statit Custom QC, use the menu selection:
  QC → Variable charts → X-bar Chart

The dialogs lead you through the creation of the chart.

If you have any difficulties, open the Help file using the **Help** button on the X-bar Chart dialog, or select **Help → Contents** and find "X-bar Chart" in the Index.

Range Chart—R-Chart

The Range chart can be combined with I charts and X-bar charts. The chart names combine the corresponding chart initials.

Range charts measure the variation in the data. An example is the weather report in the newspaper that gives the high and low temperatures each day. The difference between the high and the low is the range for that day.

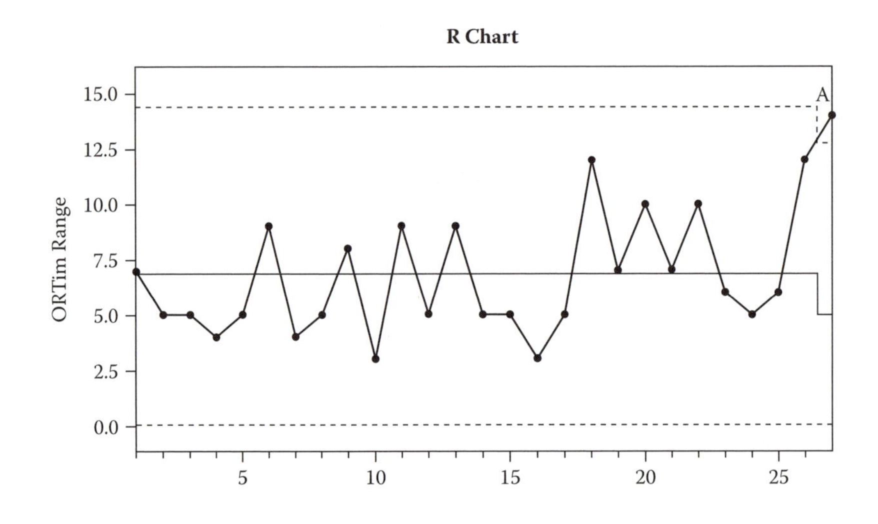

To create a Range chart:

- In Statit Express QC, this is always combined with the X-bar or I chart.
- In Statit Custom QC:
  Select: **QC → Variable charts → R Chart**

The dialogs lead you through the creation of the charts.

If you have any difficulties, open the Help file using the **Help** button on the R Chart dialog, or select **Help → Contents** and find "R Chart" in the Index.

Moving Range Chart—MR Chart

This type of chart displays the moving range of successive observations. A moving range chart can be used when it is impossible or impractical to collect more than a single data point for each subgroup.

This chart can be paired with an individual chart, which is then called an Individual Moving Range (IR) chart. An individual chart is used to highlight the changes in a variable from a central value, the mean. The moving range chart displays variability among measurements based on the difference between one data point and the next.

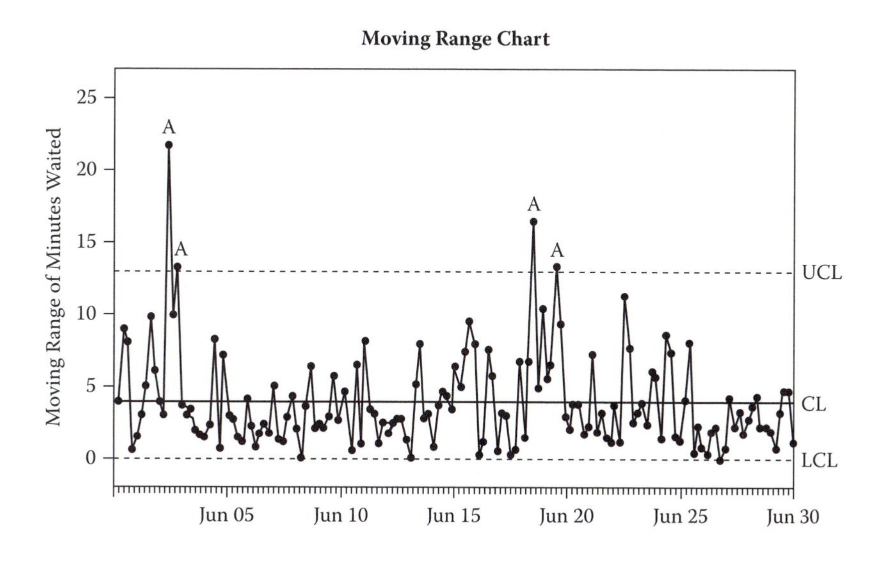

To create a Moving Range chart:

- In Statit Express QC, click on the **"MR"** button on the menu bar.
- In Statit Custom QC, use the menu selection:
  QC → Variable charts → Moving Range Chart

The dialogs lead you through the creation of the chart.

If you have any difficulties, open the Help file using the **Help** button on the Moving Range Chart dialog, or select **Help → Contents** and find "Moving Range Chart" in the Index.

Combination Charts

Individual and Range Charts—IR Charts

This pair of variable control charts is often offered together for quality control analysis. The Individual chart, the upper chart in the figure below, displays changes to the process output over time in relation to the center line which represents the mean. The Moving Range chart, the lower chart in the figure below, analyzes the variation between consecutive observations, which is a measure of process variability.

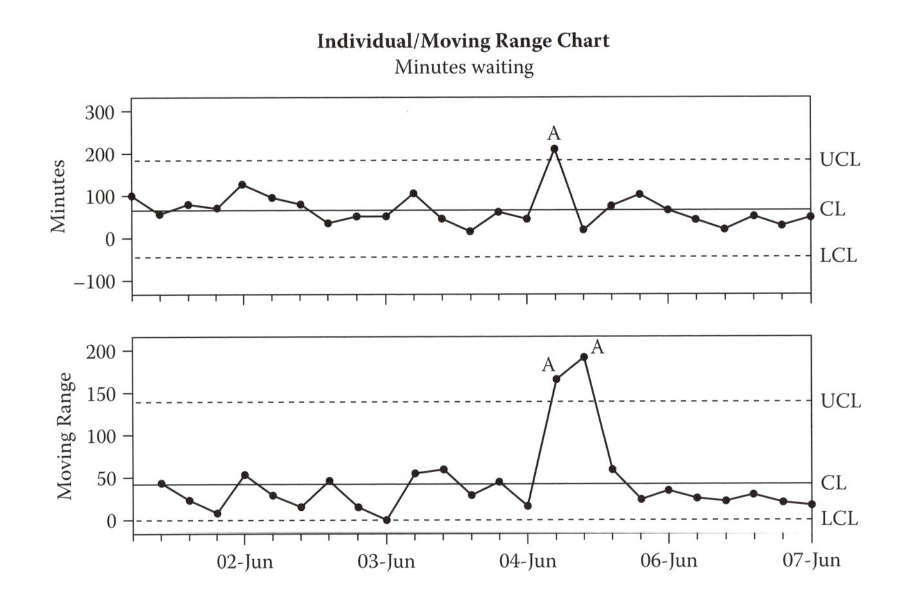

To create an Individual/Moving Range chart:

- In Statit Express QC click on the **"IR"** button on the menu bar.
- In Statit Custom QC:
  Select: **QC → Variable charts → Individual Chart**
  In the **Multiple Charts** sub dialog, select **Top of a double chart**
  Select: **QC → Variable charts → Moving Range Chart**

In the **Multiple Charts** sub dialog, select **Bottom of a double chart**
The dialogs lead you through the creation of the charts.

If you have any difficulties, open the Help file using the **Help** button on the Individual Chart or Moving Range Chart dialog, or select **Help → Contents** and find "Individual Chart" or "Moving Range Chart" in the Index.

Average & Range Charts—X-bar and R Charts

Variable and Range control charts are often displayed together for quality control analysis. The X-bar chart, the upper chart in the figure below, is a graphic representation of the variation among the subgroup averages, the R chart, the lower chart in the figure below, looks at variability within these subgroups.

The variation within subgroups is represented by the range (R). The range of values for each subgroup is plotted on the Y-axis of the R chart. The centerline is the average or mean of the range.

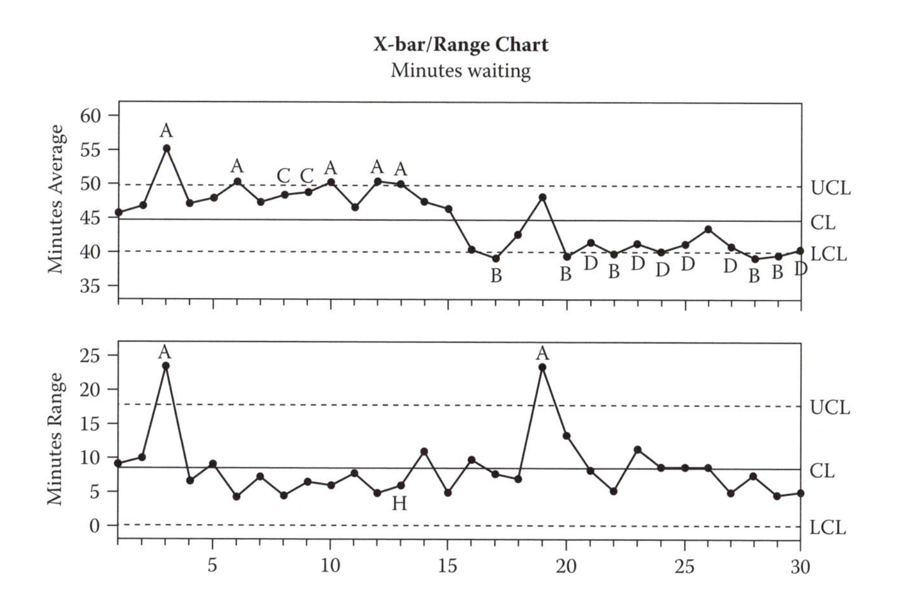

To create an X-bar/Range chart:

- In Statit Express QC click on the **"X-bar R"** button on the menu bar.
- In Statit Custom QC:
  Select: **QC → Variable charts → X-bar Chart**
  In the **Multiple Charts** sub dialog, select **Top of a double chart**
  Select: **QC → Variable charts → R Chart**

In the **Multiple Charts** sub dialog, select **Bottom of a double chart**
The dialogs lead you through the creation of the charts.

If you have any difficulties, open the Help file using the **Help** button on the X-bar Chart or R Chart dialog, or select **Help → Contents** and find "X-bar Chart" or "R Chart" in the Index.

X-bar Standard Deviation Charts—X-bar and S Charts

This pair of variable control charts is often displayed together for quality control analysis. The X-bar chart, the upper chart in the figure below, displays the variation in the means between the subgroups. The s chart, the lower chart in the figure below, looks at variability within these subgroups.

In this pair of charts, the variation within subgroups is represented by the standard deviation. The standard deviation is plotted on the y-axis, and is a measure of the spread of values for each subgroup. The centerline is the average or mean of these sub-group standard deviations.

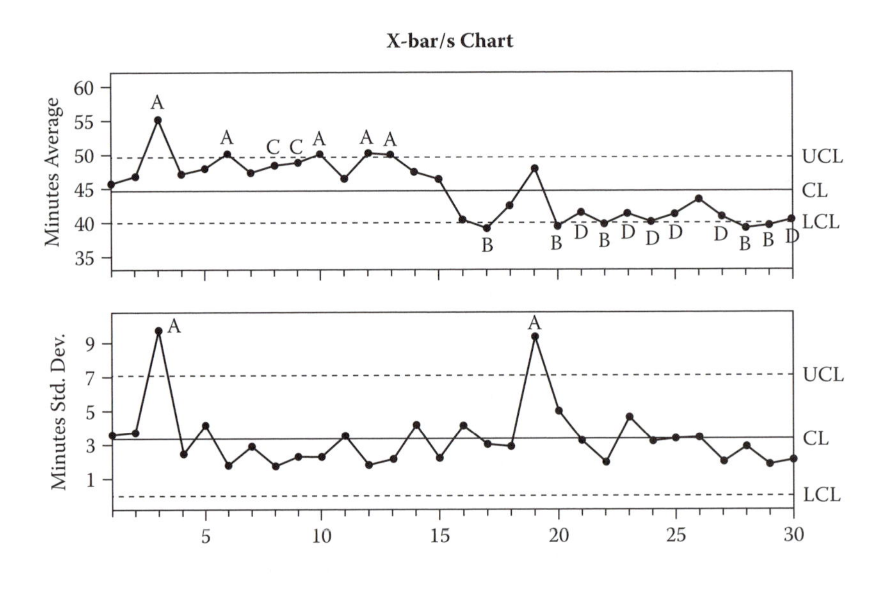

To create an X-bar/s chart:

- In Statit Express QC click on the **"X-bar S"** button on the menu bar.
- In Statit Custom QC:
  Select: **QC → Variable charts → X-bar Chart**
  In the **Multiple Charts** sub dialog, select **Top of a double chart**
  Select: **QC → Variable charts → s Chart**

  In the **Multiple Charts** sub dialog, select **Bottom of a double chart**

The dialogs lead you through the creation of the charts.

If you have any difficulties, open the Help file using the **Help** button on the X-bar Chart or s Chart dialog, or select **Help → Contents** and find "X-bar Chart" or "s Chart" in the Index.

Range vs. Standard Deviation

In Statit QC you can choose to use a standard deviation chart, the s-chart, instead of the Moving Range chart. The Range chart is often used because the standard deviation is a more accurate and therefore more difficult measurement. Now that computers are automatically calculating the standard deviation, the s-chart can be used in all situations. This is called the X-bar S chart.

A standard deviation formula is used to calculate the differences in the data. This calculation can be used in cases where the sub-group sample size is large and sampling methods support the modeling of the data as normal distribution.

Process Capability Chart—cp Chart

Process capability analysis is used to adjust the process until virtually all of the product output meets the specifications. Once the process is operating in control, capability analysis attempts to answer the question: Is the output meeting specifications, or is the process capable? If it is not, can the process be adjusted to make it capable?

The process capability chart contains a normal curve superimposed over a histogram of the data, followed by several statistics. A process is said to be capable if its output falls within the specifications virtually 100% of the time.

One goal of Statistical Process Control is to determine if specifications are in fact possible in the current process. If the following statements are true, a process capability chart can be an appropriate tool for measuring the inherent reproducibility of the process and monitoring the degree to which it can meet specifications:

- The process is stable and in control.
- The data are normally distributed.
- Specification limits fall on either side of the centerline.
- You are investigating whether your process is capable of meeting specifications.

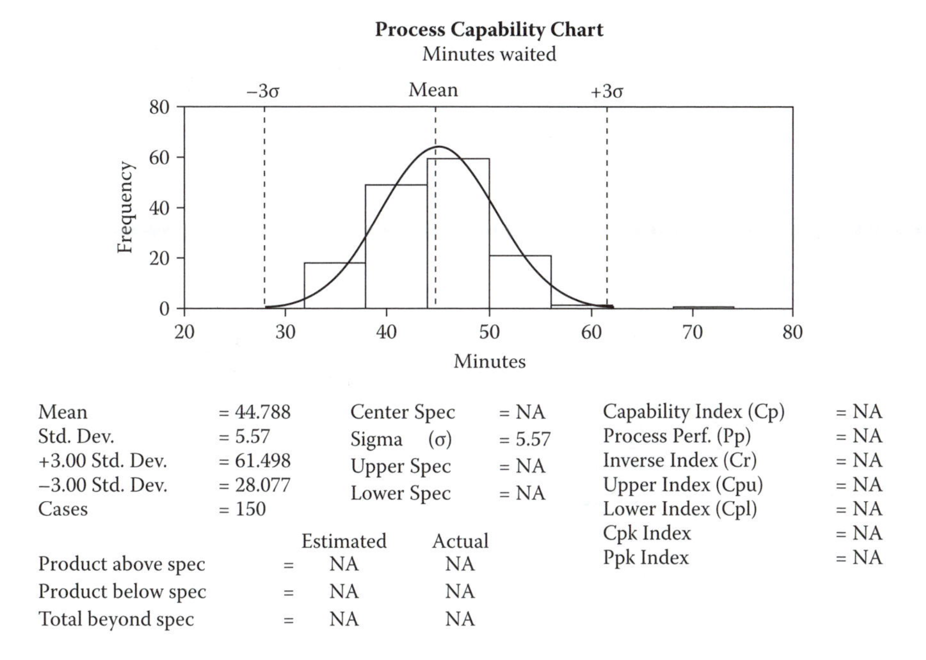

To create a Process Capability chart:

- In Statit Express QC click on the **Process Capability (cp) button** on the menu bar.
- In Statit Custom QC:
  Select: **QC** → **Variable charts** → **Process Capability Chart**

The dialogs lead you through the creation of the charts.

If you have any difficulties, open the Help file using the **Help** button on the Process Capability Chart dialog, or select **Help** → **Contents** and find "Process Capability Chart" in the Index.

Note: Specification Limits are the boundaries, or tolerances, set by management, engineers or customers which are based on product requirements or service objectives. Specification Limits are NOT established by the process itself, and may not even be possible within the given process.

Attribute Data Charts

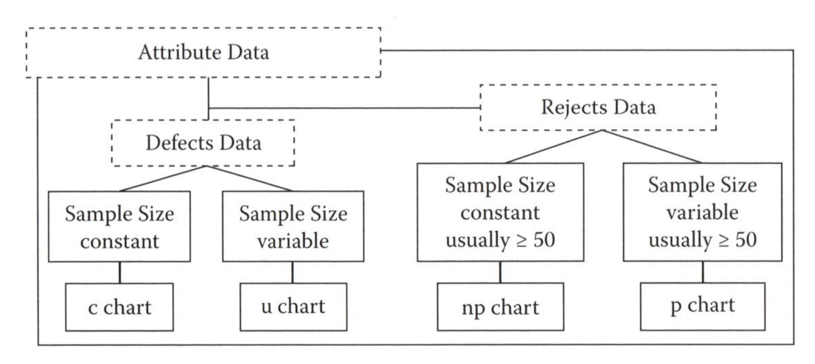

Defects Data vs. Rejects Data

Again, attribute data represents particular characteristics of a product or system that can be counted, not product measurements. They are characteristics that are present or not present. This is known as discrete data, and is measured only with whole numbers. Examples include:

Acceptable vs. non-acceptable
Forms completed with errors vs. without errors
Number of prescriptions with errors vs. without

Attribute data has another distinctive characteristic. In quality control analysis, this countable data falls into one of two categories:

- **Defects** data is the number of non-conformities within an item. There is no limit to the number of possible defects. *Defects charts* count the number of defects in the inspection unit.
- **Rejects** data where the entire item is judged to conform to product specifications or not. The count for each item is limited to 1 or 0. *Rejects charts* count the number of rejects in a subgroup.

One way to determine what type of data you have is to ask, "Can I count both the occurrences AND non-occurrences of the defective data?" For example, you can count how many forms have errors and how many do not, however you cannot count how many errors were NOT made on the form. If you can count both occurrences and non-occurrences, you have rejects data. If the non-occurrences cannot be determined, then you have defects data.

For example:

If you are counting the number of errors made on an insurance form, you have an example of the defects per form. There is no limit to the number of defects that can be counted on each form.

If you are counting the total number of forms that had one or more errors, then you have a count of the rejected units. This is either one or zero rejects per unit.

Summary of Defects vs. Rejects Data

Defects Charts

Attribute charts for cases in which the possible occurrences are infinite or cannot be counted. They count the number of non-conformities within an item.

Rejects Charts

Attribute Data charts for the cases in which rejected whole units are counted. These figures can be described as ratios instead of just counts.

Subgroup Size—Constant or Changing?

Subgroup size is another important data characteristic to consider in selecting the right type of chart. When constructing attribute control charts, a subgroup is the group of units that were inspected to obtain the number of defects or the number of rejects.

To choose the correct chart, you need to determine if the subgroup size is constant or not. If constant, for example 300 forms are processed every day, then you can look at a straight count of the defective occurrences. If the subgroup size changes, you need to look at the percentage or fraction of defective occurrences.

For example:

> An organization may have a day in which 500 insurance forms are processed and 50 have errors vs. another day in which only 150 are processed and 20 have errors. If we only look at the count of errors, 50 vs. 20, we would assume the 50 error day was worse. But when considering the total size of the subgroup, 500 vs. 150, we determine that on the first day 10% had errors while the other day 13.3% had errors.

Now that we understand the different types of attribute data, let's move on to the specific charts for analyzing them.

Attribute Charts—Defects and Rejects Charts

There are four different types of attribute charts. For each type of attribute data, defects and rejects, there is a chart for subgroups of constant size and one for subgroups of varying size.

Remember:

Defects Charts count the number of defects within the inspection unit.
Rejects Charts count the number of rejected units in a subgroup.

Defects Charts

The two defects charts are the c chart and the u chart. The **c** refers to count of defects in a subgroup of constant size. The **u** is a per unit count within a variable size subgroup.

A mnemonic to help you remember that the c chart represents Defects data is to think back to your school days and the C grade you got in a class when the number of defects or errors within one test exceeded the threshold. Another way to remember which subgroup type goes with which chart is that c is for "constant" and u is for "un-constant."

c Chart—Constant Subgroup Size

A c chart, or Count chart, is an attribute control chart that displays how the number of defects, or nonconformities, for a process or system is changing over time. The number of defects is collected for the *area of opportunity* in each subgroup. The area of opportunity can be either a group of units or just one individual unit on which defect counts are performed. The c chart is an indicator of the consistency and predictability of the level of defects in the process.

When constructing a c chart, it is important that the area of opportunity for a defect be constant from subgroup to subgroup since the chart shows the total number of defects. When the number of items tested within a subgroup changes, then a u chart should be used, since it shows the number of defects per unit rather than total defects.

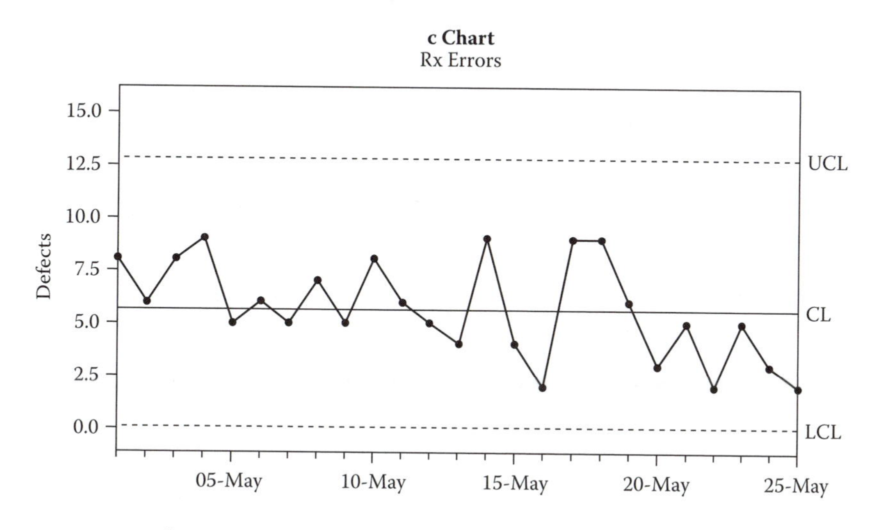

To create a C chart:

- In Statit Express QC click on the **"c"** button on the menu bar.
- In Statit Custom QC:
  Select: **QC → Attribute charts → C Chart**

The dialogs lead you through the creation of the charts.

If you have any difficulties, open the Help file using the **Help** button on the C Chart dialog, or select **Help** → **Contents** and find "C Chart" in the Index.

u Chart—Varying Subgroup Size

A u chart (u is for Unit) is an attribute control chart that displays how the frequency of defects, or nonconformities, is changing over time for a process or system. The number of defects is collected for the area of opportunity in each subgroup. The area of opportunity can be either a group of items or just one individual item on which defect counts are performed. The u chart is an indicator of the consistency and predictability of the level of defects in the process.

A u chart is appropriate when the area of opportunity for a defect varies from subgroup to subgroup. This can be seen in the shifting UCL and LCL lines that depend on the size of the subgroup. This chart shows the number of defects per unit. When the number of items tested remains the same among all the subgroups, then a c chart should be used since a c chart analyzes total defects rather than the number of defects per unit.

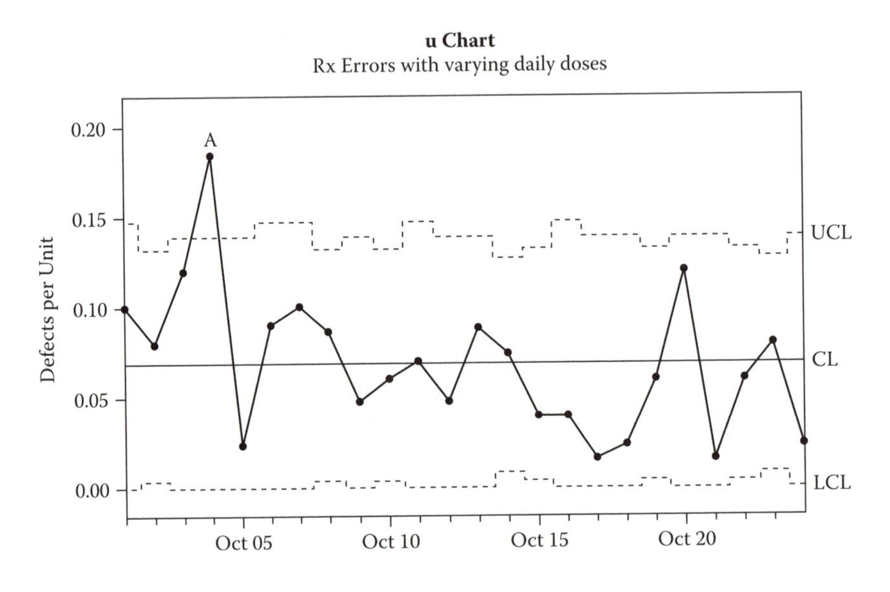

To create a U chart:

- In Statit Express QC click on the **"u"** button on the menu bar.
- In Statit Custom QC:
  Select: **QC** → **Attribute charts** → **U Chart**

The dialogs lead you through the creation of the charts.

If you have any difficulties, open the Help file using the **Help** button on the U Chart dialog, or select **Help** → **Contents** and find "U Chart" in the Index.

Rejects Charts

The two types of Rejects charts are the p chart and the np chart. The name of the p chart stands for the *P*ercentage of rejects in a subgroup. The name of the np chart stands for the *N*umber of rejects within a p-type chart. You can also remember it as "not percentage" or "not proportional."

A mnemonic to remember that the p chart and its partner the np chart represents Rejects data is to think of P as a "pea" and a canning plant that is rejecting cans of peas if they are not 100% acceptable. As p and np are a team, you should be able to recall this with the same story.

np Chart—Number of Rejects Chart for Constant Subgroup Size

An np chart is an attribute control chart that displays changes in the number of defective products, rejects or unacceptable outcomes. It is an indicator of the consistency and predictability of the level of defects in the process.

The np chart is only valid as long as your data are collected in subgroups that are the same size. When you have a variable subgroup size, a p chart should be used.

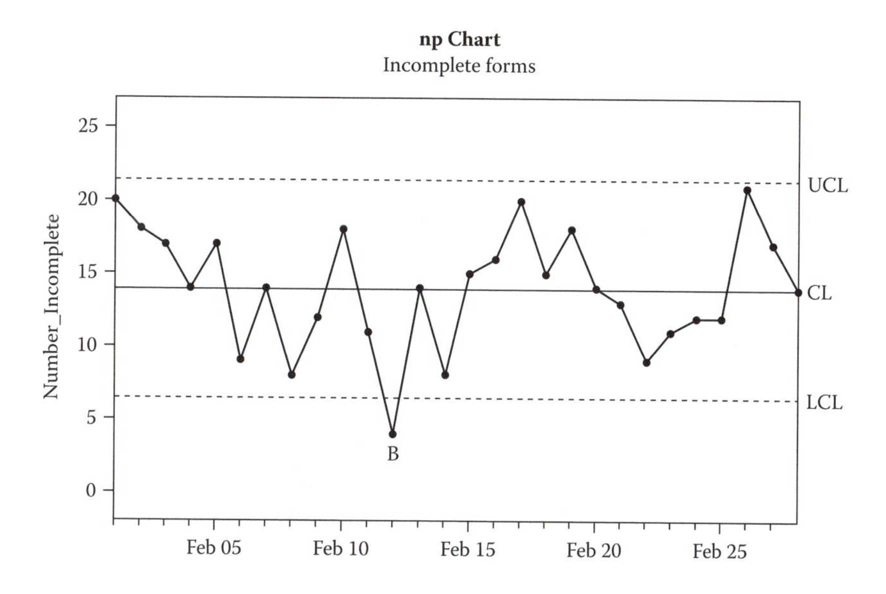

To create an NP chart:

- In Statit Express QC click on the **"np"** button on the menu bar.
- In Statit Custom QC:
  Select: **QC → Attribute charts → NP Chart**

The dialogs lead you through the creation of the charts.

If you have any difficulties, open the Help file using the **Help** button on the NP Chart dialog, or select **Help → Contents** and find "NP Chart" in the Index.

p Chart—Percentage Chart for Varying Subgroup Size

A p chart is an attribute control chart that displays changes in the proportion of defective products, rejects or unacceptable outcomes. It is an indicator of the consistency and predictability of the level of defects in the process.

Since a p chart is used when the subgroup size varies, the chart plots the proportion or fraction of items rejected, rather than the number rejected. This is indicated by the shifting UCL and LCL lines that depend on the size of the subgroup. For each subgroup, the proportion rejected is calculated as the number of rejects divided by the number of items inspected. When you have a constant subgroup size, use an np chart instead.

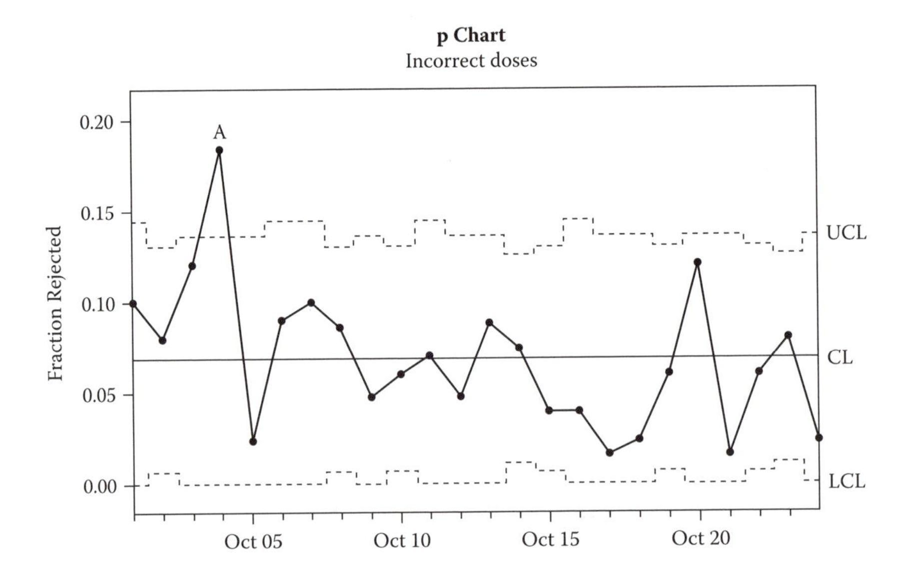

To create a P chart:

- In Statit Express QC click on the **"p"** button on the menu bar.
- In Statit Custom QC:
  Select: **QC → Attribute charts → P Chart**

The dialogs lead you through the creation of the charts.

If you have any difficulties, open the Help file using the **Help** button on the P Chart dialog, or select **Help → Contents** and find "P Chart" in the Index.

Conclusion—Time to Put It All Together ...

You now have the basic fundamentals of statistical process control. You should understand that it measures variance, and is used to determine if a process is in control. You should understand the different steps necessary in performing SPC, including problem identification, data gathering, prioritization and analysis. You should understand the key concepts and vocabulary for using SPC charts. Finally you should understand the different types of data and be able to select which control chart to use, depending on the data you have gathered.

Now that you have read and understood these concepts, let's do some hands-on work. The accompanying Technical Primer for Statit QC gives step-by-step instructions on how to use Statit Express QC to create the charts described here.

AT&T's Statistical Quality Control Standards

The rules for:

X-bar charts
Individual charts
Median charts
R charts when the minimum subgroup size is at least 4

a) 1 point above Upper Spec
b) 1 point below Lower Spec

A) 1 point above Zone A
B) 1 point below Zone A
C) 2 of 3 successive points in upper Zone A or beyond
D) 2 of 3 successive points in lower Zone A or beyond
E) 4 of 5 successive points in upper Zone B or beyond
F) 4 of 5 successive points in lower Zone B or beyond
G) 8 points in a row above centerline
H) 8 points in a row below centerline
I) 15 points in a row in Zone C (above and below center)
J) 8 points on both sides of center with 0 in Zone C
K) 14 points in a row alternating up and down
L) 6 points in a row steadily increasing or decreasing

The rules for:

R charts when the minimum subgroup size is less than 4

a) 1 point above Upper Spec
b) 1 point below Lower Spec
A) 1 point above Zone A
B) 2 successive points in or above upper Zone A
C) 3 successive points in or above upper Zone B
D) 7 successive points in or above upper Zone C
E) 10 successive points in or below lower Zone C
F) 6 successive points in or below lower Zone B
G) 4 successive points in lower Zone A

The rules for:

s charts
Moving Average charts
Moving Range charts

a) 1 point above Upper Spec
b) 1 point below Lower Spec
A) 1 point above Zone A
B) 1 point below Zone A

The rules for:

p charts
np charts
c charts
u charts

a) 1 point above Upper Spec
b) 1 point below Lower Spec
A) 1 point above Zone A
B) 1 point below Zone A
C) 9 points in a row above centerline
D) 9 points in a row below centerline
E) 6 points in a row steadily increasing or decreasing
F) 14 points in a row alternating up and down

Glossary

Accuracy: Accuracy of measurements refers to the closeness of agreement between observed values and a known reference standard. Any offset from the known standard is called bias.

Attribute data: Qualitative data that can be counted for recording and analysis. Examples include: number of defects, number of errors in a document; number of rejected items in a sample, presence of paint flaws. Attributes data are analyzed using the p-, np-, c- and u-charts.

Average: See mean.

Average Run Length (ARL): The average or expected number of points that must be plotted on a control chart before the chart signals an out-of-control condition.

Bias: The offset of a measured value from the true population value.

Binomial Distribution: A discrete probability distribution used for counting the number of successes and failures, or conforming and non-conforming units. This distribution underlies the p-chart and the np-chart.

Box and Whisker Plot: A graphical display of data that shows the median and upper and lower quartiles, along with extreme points and any outliers.

Capability: The amount of variation inherent in a stable process. Capability can be determined using data from control charts and histograms and is often quantified using the C_p and C_{pk} indices.

Cause-and-Effect Diagram: A quality control tool used to analyze potential causes of problems in a product or process. It is also called a fishbone diagram or an Ishikawa diagram after its developer.

c-Chart: A control chart based on counting the number of defects per constant size subgroup. Also known as a Count of Nonconformities chart. The c-chart is based on the Poisson distribution.

Center Line (CL): The line on the control chart that represents the long-run expected or average value of the quality characteristic that corresponds to the in-control state which occurs when only chance causes are present.

Central Limit Theorem: An important statistical theorem that states that subgroup averages tend to be normally distributed even if the output as a whole is not. This allows control charts to be widely used for process control, even if the underlying process is not normally distributed.

Common Causes: Problems with the system itself that are always present, influencing all of the production until found and removed. These are "common" to all manufacturing or production output. Also called chance causes, system causes or chronic problems. Common causes contrast to special causes.

Continuous Improvement: The ongoing improvement of products, services, or processes through incremental and breakthrough improvements.

Control Chart: A graphical mechanism for deciding whether the underlying process has changed based on sample data from the process. Control charts help determine which causes are "special" and thus should be investigated for possible correction. Control charts contain the plotted values of some statistical measure for a series of samples or subgroups, along with the upper and lower control limits for the process.

Control Limits: Statistically calculated control chart lines which indicate how the process is behaving and whether the process is in control. There is typically an upper control limit (UCL) and a lower control limit (LCL). If the process is in control and only common causes are present, nearly all of the sample points fall within the control limits. Sometimes called the Natural Process Limits for the sample size.

Correlation: A measure of the linear relation between two variables. If both variables grow larger (or smaller) together, it is called positive correlation. If one variable becomes smaller as the other grows larger, it is called negative correlation.

Count Data: See attribute data.

C_p: A measure of the capability of a process to produce output within the specifications. The measurement is made without regard to the centering of the process.

$$c_p = \frac{USL\text{-}LSL}{6\sigma}$$

C_{pk}: A measure of the capability of the process to produce output within the specifications. The centering of the process is taken into consideration by looking at the minimum of the upper specification limit capability ($C_{pu} = (USL - \bar{x})/3\sigma$) and the lower specification limit capability ($C_{pl} = (\bar{x} - LSL)/3\sigma$); $C_{pk} = \min(C_{pu}, C_{pl})$

CUSUM: A control chart designed to detect small process shifts by looking at the Cumulative SUMs of the deviations of successive samples from a target value.

Design of Experiments: A branch of applied statistics dealing with planning, conducting, analyzing, and interpreting controlled tests which are used to identify and evaluate the factors that control a value of a parameter of interest.

Distribution: A mathematical model that relates the value of a variable with the probability of the occurrence of that value in the population.

EWMA charts: An Exponentially Weighted Moving Average control chart that uses current and historical data to detect small changes in the process. Typically, the most recent data is given the most weight, and progressively smaller weights are given to older data.

Histogram: A graph of the observed frequencies versus each value or range of values for a set of data. A histogram provides a graphical summary of the variation in the data.

Hypothesis Testing: A procedure that is used on a sample from a population to investigate the applicability of an assertion (inference) to the entire population. Hypothesis testing can also be used to test assertions about multiple populations using multiple samples.

In-Control Process: A process in which the quality characteristic being evaluated is in a state of statistical control. This means that the variation[s] among the observed samples can all be attributed to common causes, and that no special causes are influencing the process.

Individual: A single unit or a single measurement of a quality characteristic, usually denoted as X. This measurement is analyzed using an individuals chart, CUSUM or EWMA chart.

Individuals Chart: A control chart for processes in which individual measurements of the process are plotted for analysis. Also called an I-chart or X-chart.

Mean: A measure of the location or center of data. Also called the average. The mean is calculated by summing all of the observations and dividing by the number of observations.

$$\text{Sample mean} = \overline{X} = \frac{\sum_{i-1}^{n} X_1}{n} = \frac{X_1 + X_2 + \ldots X_n}{n}$$

Median: The "middle" value of a group of observations, or the average of the two middle values. The median is denoted by x tilde (*X*%).

Mixing: A generally improper sampling technique that arises in practice when the output from several processes is first thoroughly mixed and then random samples are drawn from the mixture. This may increase the sample variability and make the control chart less sensitive to process changes. This action violates the fundamental rule of rational sampling.

Mode: The observation that occurs most frequently in a sample. The data can have no mode, be unimodal, bimodal, etc.

Moving Range: A measure used to help calculate the variance of a population based on differences in consecutive data. Two consecutive individual data values are compared and the absolute value of their difference is recorded on the moving range chart. The moving range chart is typically used with an Individuals (X) chart for single measurements.

Nonconforming Unit: A unit with one or more nonconformities or defects. Also called a reject.

Nonconformity: A specified requirement that is not fulfilled, such as a blemish, defect or imperfection.

Normal Distribution: A continuous, symmetrical, bell-shaped frequency distribution for variables data that is the basis for control charts for variables, such as x-bar and individuals charts. For normally distributed values, 99.73% of the population lies within ± 3 standard deviations of the mean. According to the Central Limit Theorem, subgroup averages tend to be normally distributed even if the output as a whole is not.

np-Chart: A control chart based on counting the number of defective units in each constant size subgroup. The np-chart is based on the binomial distribution.

Outliers: Unusually large or small observations relative to the rest of the data.

Over control: An element often introduced into a process by a well-meaning operator or controller who considers any appreciable deviation from the target value as a special cause. In this case, the operator is wrongly viewing common-cause variation as a fault in the process. Over control of a process can actually increase the variability of the process and is viewed as a form of tampering.

Pareto Chart: A problem-solving tool that involves ranking all potential problem areas or sources of variation according to their contribution to cost or total variation. Typically, 80% of the effects come from 20% of the possible causes, so efforts are best spent on these "vital few" causes, temporarily ignoring the "trivial many" causes.

p-Chart: A control chart based on the proportion of nonconforming units per subgroup. The p-chart is based on the binomial distribution.

PDCA: The Plan-Do-Check-Act cycle is a four-step process for quality improvement.

Plan: A plan for improvement is developed.

Do: Experiments are initiated to test the feasibility of the plan, or the plan is carried out.

Check (or Study): The effects of the plan are observed.

Act: The results are studied to see what was learned and what can be predicted.

The cycle continues until the desired improvement is realized

Percentiles: Percentiles divide the ordered data into 100 equal groups. The kth percentile p_k is a value such that at least k% of the observations are at or below this value and (100 - k)% of the observations are at or above this value.

Poisson Distribution: A probability distribution used to count the number of occurrences of relatively rare events. The Poisson distribution is used in constructing the c-chart and the u-chart.

Precision: Precision of measurements refers to their long-run variation (s^2). It is a measure of the closeness between several individual readings.

Process Capability: A measure of the ability of a process to produce output that meets the process specifications.

Quartile: Quartiles divide the ordered data into 4 equal groups. The second quartile (Q2) is the median of the data.

Random Sampling: A subset of the population chosen such that each member of the population has an equal probability of being included in the sample.

Range: A measure of the spread of the data, calculated as highest value minus lowest value. Range = $R = X_{max} - X_{min}$

Rational Subgroups: A principle of sampling which states that the variation between subgroups or samples should be solely attributable to the common causes in the system rather than the sampling method. Rational subgroups are usually chosen so that the variation represented within each subgroup is as small as feasible for the process, so that any changes in the process, or special causes, appear as differences between subgroups. Rational subgroups are typically made up of consecutive pieces, although random samples are sometimes used.

R-Chart: A control chart based on the range (R) of a subgroup, typically used in conjunction with an x-bar chart.

Run: A consecutive number of points consistently increasing or decreasing, or above or below the centerline. A run can be evidence of the existence of special causes of variation that should be investigated.

Runs Chart: A simple graphic representation of a characteristic of a process which shows plotted values of some statistic gathered from the process. The graphic can be analyzed for trends or other unusual patterns.

S-Chart: A control chart based on the standard deviation, s, of a subgroup. The s-chart is typically used in conjunction with an x-bar chart.

Sample: A subset of data from a population that can be analyzed to make inferences about the entire population.

Sampling Distribution: The probability distribution of a statistic. Common sampling distributions include t, chi-square (χ^2) and F.

Scatter Plots: A graphical technique used to visually analyze the relationship between two variables. Two sets of data are plotted on a graph, with the y-axis being used for the variable to be predicted and the x-axis being used for the variable to make the prediction.

Sensitizing Rules: Control chart interpretation rules that are designed to increase the responsiveness of a control chart to out-of-control conditions by looking for patterns of points that would rarely happen if the process has not changed.

Short-run Techniques: Adaptations made to control charts to help determine meaningful control limits in situations when only a limited number of parts are produced or when a limited number of services are performed. Short-run techniques usually look at the deviation of a quality characteristic from a target value.

Six Sigma: A high-performance, data-driven approach to analyzing the root causes of business problems and solving them. Six-sigma techniques were championed by Motorola.

Skewness: The tendency of the data distribution to be non-symmetrical. Negative skewness denotes more small observations, while positive skewness denotes more large observations. Skewed data may affect the validity of control charts and other statistical tests based on the normal distribution.

Special Causes: Causes of variation which arise periodically in a somewhat unpredictable fashion. Also called assignable causes, local faults, or sporadic problems. Contrast to common causes. The presence of special causes indicates an out-of-control process.

Specification: The written or engineering requirements for judging the acceptability of the output of a process.

Spread: The amount of variability in a sample or population.

Stability: A process is considered stable if it is free from the influences of special causes. A stable process is said to be in control.

Standard Deviation: A measure of the spread of a set of data from its mean, abbreviated: σ for a population s for a sample. The standard deviation is the square root of the variance.

Statistic: A value calculated from or based on sample data which is used to make inferences about the population from which the sample came. Sample mean, median, range, variance and standard deviation are commonly calculated statistics.

Statistical Control: The condition describing a process from which all special causes of variation have been removed and only common causes remain.

Statistical Process Control (SPC): A collection of problem solving tools useful in achieving process stability and improving capability through the reduction of variability. SPC includes using control charts to analyze a process to identify appropriate actions that can be taken to achieve and maintain a state of statistical control and to improve the capability of the process.

Statistical Quality Control (SQC): Another name commonly used to describe statistical process control techniques.

Stratified Sampling: Stratification arises in practice when samples are collected by drawing from each of several processes, for example machines, filling heads or spindles. Stratified sampling can increase the variability of the sample data and make the resulting control chart less sensitive to changes in the process.

Subgroup: Another name for a sample from the population.

Tampering: An action taken to compensate for variation within the control limits of a stable system. Tampering increases rather than decreases variation, as in the case of over control.

Type I Error: Occurs when a true hypothesis about the population is incorrectly rejected. Also called false alarm. The probability of a Type I error occurring is designated by α.

Type II Error: Occurs when a false hypothesis about the population is incorrectly accepted. Also called lack of alarm. The probability of a Type II error occurring is designated by β.

u-Chart: A control chart based on counting the number of nonconformities or defects per inspection unit. The u-chart is based on the Poisson distribution.

Variables Data: Data values which are measurements of some quality or characteristic of the process. The data values are used to construct the control charts. This qualitative data is used for the x-bar, R-, s- and individuals charts, as well as the CUSUM and moving range charts.

Variance: A measure of the spread of a set of data from its mean, abbreviated:σ^2 for a population s^2 for a sample

$$\text{Sample Variance} = s^2 = \frac{\sum_{i=1}^{n}(X_i - \bar{X})^2}{n-1} = \frac{\sum_{i=1}^{n} X_i^2 - \frac{\left(\sum_{i=1}^{n} X_i\right)^2}{n}}{n-1}$$

Variation: The differences among individual results or output of a machine or process. Variation is classified in two ways: variation due to common causes and variation due to special causes.

X-Chart: A control chart used for process in which individual measurements of the process are plotted for analysis. Also called an individuals chart or I-chart.

X-bar ($\bar{x}$) Chart: A control chart used for processes in which the averages of subgroups of process data are plotted for analysis.

Bibliography

Nelson, Loyd S. (1985), "Interpreting Shewhart X Control Charts," *Journal of Quality Technology*, 17:114-16.

Steel, R. G. D. and J. H. Torrie (1980), *Principles and Procedures of Statistics*. New York: McGraw-Hill.

Western Electric Company (1956), *Statistical Quality Control Handbook*, available from ATT Technologies, Commercial Sales Clerk, Select Code 700-444, P.O. Box 19901, Indianapolis, IN 46219, 1-800-432-6600.

URL

Statit Software: http://www.statit.com

Index

D

P

T

U

X

Z